Ocean Energy

Energy from wave and tidal power is a key component of current policies for renewable sources of energy. This book provides the first comprehensive exploration of legal, economic, and social issues related to the emerging ocean energy industry, in particular wave and tidal energy technologies.

This industry is rapidly developing, and considerable technical literature has developed around the technology. However, it is shown that challenges relating to regulation and policy are major impediments to industry development, and these aspects have not previously been sufficiently highlighted and studied. The book informs policymakers, industry participants, and researchers of the key issues in this developing field. Ocean energy is considered in the context of the blue economy and an industrialising ocean, and the topics covered include development of policy (policy instruments, risk and delay in technology development); legal aspects (consenting processes, resource management, impact assessment); human interactions (conflicts, consultation, community benefits); and spatial planning of the marine environment.

While offshore wind energy, sited in the oceans but not strictly derived from the ocean, is not the primary focus of the book, there is also discussion of the similarities and differences between offshore wind and wave and tidal power policy dimensions.

Glen Wright is a Research Fellow at the Institute for Sustainable Development and International Relations (IDDRI, SciencesPo), Paris, France.

Sandy Kerr is Director and Associate Professor, International Centre for Island Technology, Heriot-Watt University, based in the Orkney Islands, UK.

Kate Johnson is an Assistant Professor, International Centre for Island Technology, Heriot-Watt University, based in the Orkney Islands, UK.

Earthscan Oceans

For further details please visit the series page on the Routledge website: www.routledge.com/books/series/ECOCE

Ocean Energy

Governance Challenges for Wave and Tidal Stream Technologies

Edited by Glen Wright, Sandy Kerr and Kate Johnson

LONDON AND NEW YORK

from Routledge

First published 2018 by Routledge

2 Park Square, Milton Park, Abingdon, Oxfordshire OX14 4RN
52 Vanderbilt Avenue, New York, NY 10017

Routledge is an imprint of the Taylor & Francis Group, an informa business

First issued in paperback 2019

British Library Cataloguing-in-Publication Data
A catalogue record for this book is available from the British Library

Library of Congress Cataloging-in-Publication Data
A catalog record for this book has been requested

ISBN: 978-1-138-66852-2 (hbk)
ISBN: 978-0-367-40381-2 (pbk)

Typeset in Goudy
by Apex CoVantage, LLC

Contents

Tables

Figures

Contributors

Maria Campbell received her doctorate from the School of Marine Science and Engineering, University of Plymouth, UK.

Jordan Carlson is a graduate student and teaching assistant at the School for Resource and Environmental Studies, Dalhousie University, Canada.

John Colton is a Professor and the Graduate Program Coordinator of the Department of Community Development at Acadia University, Canada.

Flaxen Conway is a Professor of Sociology in the School of Public Policy, College of Liberal Arts, at Oregon State University, USA.

Patrick Devine-Wright is a Professor in Human Geography at the University of Exeter, UK.

Andrew B. Gill is Director of the Cape Eleuthera Institute, the Bahamas, and visiting researcher within the Offshore Renewable Energy and Engineering Centre at Cranfield University, UK.

Jiska de Groot is the Leader of the Energy, Poverty and Development Group at the University of Cape Town, South Africa.

Kate Johnson is an Assistant Professor, International Centre for Island Technology, Heriot-Watt University, based in the Orkney Islands, UK.

Simon Jude is a Senior Lecturer at the Cranfield Institute for Resilient Futures, Cranfield University, UK.

Sandy Kerr is Director and Associate Professor, International Centre for Island Technology, Heriot-Watt University, based in the Orkney Islands, UK.

Shelley MacDougall is a Professor Emeritus (Finance) of the F.C. Manning School of Business Administration and the Acadia Tidal Energy Institute at Acadia University, Nova Scotia, Canada.

Craig Mauelshagen is the Head of Research: Risk Analytics at Business Modelling Associates, UK.

Anne Marie O'Hagan is a Senior Research Fellow at the Environmental Research Institute, University College Cork, Ireland.

Kieran Reilly obtained his PhD from the University College Cork, Ireland, and is now working in the Irish Marine Institute.

Laura Watts is Associate Professor in the Technologies in Practice Group at IT University of Copenhagen, Denmark.

Stephanie Weir is a PhD candidate at Heriot-Watt University, UK.

Edward Willsteed is a PhD candidate at Cranfield University, UK.

Bouke Wiersma obtained his doctorate from the University of Exeter, UK, and now works at the Netherlands Ministry of Infrastructure and the Environment.

Brit Ross Winthereik is Associate Professor in the Technologies Practice Group at the IT University of Copenhagen, Denmark.

Glen Wright is a Research Fellow at the Institute for Sustainable Development and International Relations (IDDRI, SciencesPo), Paris, France, and a PhD candidate at the Australian National University, Australia.

Preface

The generation of clean electricity from the ocean's waves and tides has long been a dream. The existence and extent of the resource are without question, yet ocean energy has proven incredibly hard to realise. For decades, a wild and unforgiving sea seemed to place ocean energy just beyond our grasp.

However, these technologies are now tantalisingly close to commercial deployment. The need for clean energy is urgent, and our maritime technologies have advanced in leaps and bounds. Paradoxically, the experience gained from exploiting offshore fossil fuels has helped us overcome many of the technical challenges of working in a hostile marine environment, and ocean energy is approaching a tipping point as arrays of devices begin to enter the water.

At the same time, as the push towards a 'blue economy' takes root and the seas become ever busier, the ocean energy sector now faces a new series of non-technical challenges: questions of governance of the marine commons, as well as ownership, planning, regulation, and questions regarding social, economic and cultural impacts.

This book is the first comprehensive exploration of these wide-ranging governance challenges, intended to be an essential guide for developers, regulators, stakeholders, students and all those with an interest in realising the dream of clean energy from waves and tides.

Chapter 1

Introduction

Context, technology and governance

Glen Wright, Sandy Kerr and Kate Johnson

> *Most things become exhausted with promiscuous use.*
> *This is not the case with the sea.*
>
> —Grotius (1608)

1 The industrial revolution of the oceans

At the time Hugo Grotius was formulating the Freedom of the Seas principle in 1608, the sea, vast and teeming with life, would have indeed seemed inexhaustible. Yet 400 years later, our seas are looking increasingly exhausted, and there is a growing awareness of the fragility of our blue planet. Our ever growing population and appetite for resources have driven huge increases in traditional maritime activities like fishing and shipping, while a new industrial revolution of the oceans is driving unprecedented exploitation of the marine environment, placing further pressure on ocean ecosystems and changing the way we think about marine governance.

The seas and coasts have long been strong drivers of economies worldwide, and coastal communities and ports have traditionally been hubs for ideas and innovation due to their outward-looking geography.[1] However, the potential for industrial activity and innovation in the marine environment has grown exponentially in recent decades due to three main factors. Firstly, rapid technological progress has opened up new possibilities for the exploration and exploitation of marine areas. Secondly, land and freshwater resources are finite and are being rapidly depleted. Thirdly, the need to mitigate climate change by reducing greenhouse gas emissions has increased interest in sustainable development of the marine environment.

While the industrial revolution on land precipitated the climate change era, the industrial revolution in the oceans has the potential to be part of the solution, if managed appropriately. Well managed aquaculture could provide a much needed source of food to coastal communities while preserving natural ecosystems,[2] carbon capture and storage could remove carbon from the atmosphere and store it offshore,[3] sustainable tourism can lead to improved resource management,[4] and ocean energy technologies could use the movement of the waves and tides to generate low carbon electricity.

2 Climate change and renewable energy

There is now a widely accepted scientific and political consensus that anthropogenic greenhouse gas emissions, primarily from the burning of fossil fuels, are responsible for increasing the amount of carbon dioxide in the atmosphere and causing global warming. At the Paris climate conference in 2016, world leaders negotiated an ambitious agreement to strengthen the global response to climate change, including by:

> Holding the increase in the global average temperature to well below 2°C above pre-industrial levels and to pursue efforts to limit the temperature increase to 1.5°C above pre-industrial levels, recognizing that this would significantly reduce the risks and impacts of climate change.

The majority of energy used worldwide comes from non-renewable sources, and meeting the Paris Agreement will require the rapid decarbonisation of our energy system. The Intergovernmental Panel on Climate Change has reaffirmed the role of renewable energy in confronting these challenges, stating that renewables have a "large potential to mitigate climate change" and can "contribute to social and economic development, energy access, energy security, and reduce negative impacts on the environment and health".[5]

The development of renewable energy technologies and their integration into the energy mix will reduce emissions, conserve finite resources and increase energy security, in addition to bringing additional co-benefits, such as job creation. As a result, investment in renewable energy has grown dramatically in the last decade.[6]

3 Ocean energy technologies

The imperative to decarbonise the energy system initially drove the development of solar and onshore wind power, but interest has now spread offshore.[7] The marine environment contains a number of potential sources of renewable energy, including wave and tidal energy (in this book, collectively termed 'ocean energy'), ocean thermal energy,[8] salinity gradient or osmotic energy[9] and offshore wind energy. Offshore wind in particular has seen rapid growth in recent years,[10] while wave and tidal technologies are now also beginning to attract considerable interest and investment.[11]

The oscillatory or circular motions of waves produced near the sea's surface can be converted into electricity,[12] while tidal or marine hydrokinetic energy comprises ocean currents or tides that can be directly extracted and converted into electricity.

There are a seemingly infinite number of ways that ocean energy technologies can harness the wave and tidal power of the oceans. The following quote from

Stephen Salter, inventor of the Salter's (Edinburgh) duck device for extracting wave energy, is indicative:[13]

> Waves [are] only one of many possible sources and there are many possible ways in which waves can be harnessed. There are floats, flaps, ramps, funnels, cylinders, air-bags and liquid pistons. Devices can be at the surface, the sea bed or anywhere between. They can face backwards, forwards, sideways or obliquely and move in heave, surge, sway, pitch and roll. They can use oil, air, water, steam, gearing or electro-magnets for generation. They make a range of different demands on attachments to the sea bed and connections of power cables.

Ocean energy technologies were first explored in a number of countries as a result of the 1970s oil crisis. During this time, the UK's Department of Energy ran a wave energy programme aimed at upscaling prospective devices, while the United States focused on OTEC (ocean thermal energy conversion). However, this initial enthusiasm diminished in the face of high costs and the easing of the oil crisis, and subsequent support for ocean energy was sporadic. By the late 1980s, research activity had greatly diminished, and R&D funding was insignificant through much of the 1990s.[14]

Climate change and energy concerns have brought a 'second wave' of interest to ocean energy, particularly in the UK, Ireland, Portugal, Denmark, France, Australia, South Korea, Canada, and the United States. Large-scale utilities, energy agencies and industrial companies are making significant investments in the sector,[15] and the European Marine Energy Centre (EMEC), a test bed for emerging ocean energy technologies, is fully subscribed with 14 full-scale devices generating to the grid. Similar testing centres are under development elsewhere. The military has also shown interest in supporting the development of ocean energy technologies, with a US naval base in Hawaii developing a wave energy generation project[16] and a naval base in Western Australia signing an agreement to meet its power needs through ocean energy.[17]

In August 2016, the first array of tidal turbines sent energy to the grid in the Shetland Islands,[18] while in September the Orkney Islands saw the launch of the first of four 1.5-MW tidal turbines, the first phase of a project eventually intended to reach an installed capacity of 398 MW.[19]

Devices are advancing rapidly, costs are coming down, and commercialisation appears to be on the horizon: "The technologies have been optimized, extensively tested and pilot wave energy projects have been realized. The knowledge and experiences gained lead to a development status that is ready for the market".[20]

While recent developments in ocean energy are encouraging, a word of caution is advisable. Despite successful prototype deployments and the arrival of the first arrays and large-scale projects, ocean energy remains the least developed of the renewable energy technologies. Overall, ocean energy is perhaps at the same level of development as wind technology in the 1970s and early 1980s, when a range of wind turbine concepts were being researched and developed but uncertainty remained as to which concepts would become cost-competitive. Like many new industries, ocean energy proponents have often been overly ambitious or optimistic in their assessments of the future of ocean energy, and they risk overselling the potential or pace of the industry. At the same time, a wave of recent bankruptcies and technology failures, though perhaps inevitable, have shaken the industry and cast doubt on its ability to deliver.

A similar history plagued OTEC. There was a surge of interest in in the 1970s, but the predictions did not come to fruition as interest waned. In 1980, it was thought that research in key countries had advanced the technology to a stage where commercial OTEC devices could be deployed before the end of the decade and that, by the turn of the century, OTEC would be a commercial source of energy. Hearings before the House Subcommittee on Oceanography in the United States in 1978 estimated that OTEC generation would produce 3% of US energy requirements by 2000.[21] In spite of considerable R&D efforts into the technology and the efforts of academics and legislators to improve the regulatory regime, OTEC did not ultimately live up to these expectations. While OTEC's demise was largely due to the easing of the oil crisis, in contrast to the more intractable climate and energy issues now faced by policymakers, this nonetheless cautions against making unduly optimistic assessments of the potential of ocean energy technologies.

4 Marine governance

Ocean energy devices enter into a complex legal and policy environment, as our approaches to marine governance are evolving rapidly along with our increased use of the ocean. Within their own waters, states have traditionally managed marine activities on a single-sector basis. This was somewhat functional where uses of the oceans were limited and conflicts were few. However, this paradigm has severe limitations when marine activities increase, conflicts between users become more common, and the environment is put under increasing pressure.

This period of change has generated repeated calls for a reimagining of our marine governance systems:

> As various ocean uses and the contradictions among them intensify . . . it seems likely that the oceans will become a site for imagining and creating future social institutions and relations, for land as well as for sea.[22]

> We need to get better at planning for a sustainable ocean . . . In every challenge there is an opportunity, and the opportunity here represents a new frontier in our management approach for our oceans.[23]

> We need to articulate a new discourse on sea tenure (ocean rights and responsibilities) in the twenty-first century, a new way to allocate ocean

space and marine resources without carving the oceans into private fiefdoms. As demands for ocean resources and space multiply, we need a way for private enterprise to pioneer wind, wave, and tidal-energy offshore as well as open-water aquaculture. We need new governance systems that protect the rights of this and future generations.[24]

We now face a pressing need to develop new legal and regulatory structures to address the questions raised by the plethora of unique new projects aiming to exploit our coastal waters. Increased industrialisation of the oceans has led to a number of new discourses in this area that are reshaping social and legal institutions for the sea. Governments are already in the process of restructuring the rights and rules of the oceans, and lawmakers are taking up new regulatory models and innovations in their modernisation of marine governance, such as marine spatial planning.

This new period in the evolution of marine governance has seen a move away from a focus on single industries, single objectives and particular mechanisms towards discussion of rights and ownership in the marine environment, resource management, regulation of environmental interactions of marine uses and planning of marine spaces.[25]

5 Ocean energy and marine governance: the literature

As with other novel offshore activities, ocean energy is bringing its own unique challenges to marine governance frameworks.[26] Ocean energy is not simply a technically challenging extension of onshore renewable energy; rather, the "policy environment, governance, patterns of resource use, conservation values, and distribution of ownership rights are all substantively different from the situation onshore".[27]

In the 1970s, legal research on ocean energy developed alongside new technological developments, largely focusing on the then-frontrunner technology, OTEC, identifying the relevant legal frameworks and considering how they might be improved.[28] Many issues raised by this early scholarship still ring true some 40 years later. For example, concerns in 1976 regarding the "adverse effects of the late blooming jurisdictional and environmental impediments to implementation of new technologies" are equally relevant today, as rapid technological advancements threaten to outstrip the evolution of the associated legal and regulatory frameworks.[29]

The second wave of interest in ocean energy has sparked renewed academic interest in the governance and policy issues surrounding these technologies. This interest focused initially on the technology itself but has expanded to cover issues such as community acceptance;[30] attitudes of fishers towards ocean energy development,[31] coexistence with fisheries,[32] and the economic impacts of industry development.[33] The International Network for Social Studies in Marine Energy (ISSMER) was convened in 2012,[34] and a research agenda for social studies focused on offshore renewables was developed.[35]

Legal and regulatory frameworks affect the development of new technologies and industries, and there are strong links between these issues and the issues highlighted by the social sciences research agenda.[36] A range of scholarship has begun to emerge considering such issues, including discussion of international law,[37] consenting issues,[38] environmental impact assessment,[39] and marine spatial planning.[40] A legal research agenda for ocean energy has also been developed, setting out a guiding framework for the exploration of legal and governance issues relating to ocean energy, as well as identifying some key questions for further research.[41]

Ocean energy raises a range of important policy questions and challenges all aspects of existing marine governance frameworks. In most cases, developers desire or require rights and tenure in the ocean space they seek to exploit, challenging existing conceptions of rights and ownership. Resources are limited, and there will be competition over access, challenging regulatory authorities to develop appropriate processes for resource allocation and management. The environmental impacts remain uncertain, yet deployment of devices is necessary to advance the industry, thereby challenging risk-averse regulators to provide flexibility in environmental impact assessment (EIA) processes. Finally, ocean energy arrays will have large space requirements that are unprecedented in the nearshore environment, challenging single-sector management regimes and necessitating better integration of innovative industries into marine spatial planning (MSP) efforts.

6 Contents

This book consists of 13 substantive chapters, exploring the governance challenges for wave and tidal energy technologies. Chapters 2 and 3 consider some of the economic aspects of ocean energy technology development: Chapter 2 provides an introduction to the concept of risk in the context of technology development, considering how this applies to ocean energy and how investment can be de-risked; Chapter 3 explores the value to developers of delaying investment and projects, using the Black–Scholes model to assess such behaviour in the ocean energy context.

Chapters 4 to 6 take a wide perspective on the governance aspects of ocean energy. Chapter 4 provides a detailed introduction to the development of law and governance in the marine environment, while Chapter 5 introduces the concept of marine planning and discusses the place of ocean energy within such overarching governance processes. Chapter 6 discusses rights and ownership in the marine environment, including the potential for coexistence between the quasi-private rights required by ocean energy developers and the rights of the public and other ocean users.

Chapters 7 to 9 with the project-level regulatory frameworks for consenting and environmental assessment. Chapter 7 looks at the key issues and challenges for consenting ocean energy projects and provides a discussion of reforms undertaken in the UK. This detailed discussion is supplemented by

Chapter 8, which provides an overview of consenting processes in a number of jurisdictions. Chapter 9 considers the role of environmental impact assessment processes.

Chapters 10 to 13 focus on the community and consultation aspects of ocean energy development. Chapter 10 provides an overview of the potential conflicts between ocean energy and existing uses, focusing on the fishing industry as a key flashpoint and providing pragmatic paths forward for managing conflicts. Chapter 11 introduces the concept of community benefit and how this has been discussed in the ocean energy context, and explores the balance of power and dynamics in energy policy through case studies and a comparison between onshore and offshore wind. Chapter 12 considers consultation in general, providing a number of case studies from around the world and highlighting some pathways to improved consultation processes. Finally, Chapter 13 offers a unique perspective on these community issues, exploring the intersection between ocean energy and the geographic periphery, drawing upon ethnography and science and technology studies to suggest different methods for communicating with communities and stakeholders about ocean energy.

Taken together, these chapters form a detailed and challenging exploration of the key issues confronting the development of this exciting new maritime industry. It is hoped that this book will provide a basis for further research and discussion of this emerging field at the crossroads of innovation, renewable energy and marine governance.

Notes

1 European Commission Maritime Affairs (2012) *Blue Growth: Opportunities for Marine and Maritime Sustainable Growth.* Communication from the Commission to the European Parliament, the Council, the European Economic and Social Committee and the Committee of the Regions, COM(2012) 494.
2 For example, the 23rd session of the UN FAO Committee on Fisheries noted the "increasingly important role of inland capture fisheries and aquaculture in fish production and human nutrition and emphasized the benefits of integrated resources management, the need to combat adverse impacts on the environment and for enhanced cooperation between fishers, government agencies and other stakeholders". Benedict, S. (1999) *COFI – Committee on Fisheries – 23rd Session (FI-701-23).* Food and Agriculture Organization of the United Nations.
3 Offshore Technology (2010) *Pipe Dream: Carbon Capture and Storage.* Available at: www.offshore-technology.com/features/feature94802/
4 United Nations World Summit on Sustainable Development (2002) *Sustainable Tourism – Turning the Tide.* Economic Briefing No. 4. Available at: www.earthsummit 2002.org/es/issues/tourism/tourism.pdf
5 Edenhofer, O., et al. (2011) *IPCC Special Report on Renewable Energy Sources and Climate Change Mitigation; Summary for Policy Makers.* Cambridge: Cambridge University Press.
6 For up-to-date figures on the renewable energy investment and deployment, see the latest version of the REN21 Renewables Global Status Report. Available at: www. ren21.net/ren21activities/globalstatusreport.aspx
7 Ibid.

8   Ocean Thermal Energy Conversion (OTEC) technology exploits the potential energy contained in the temperature difference between surface and deep water (thermal gradient).

9   This refers to the difference in pressure between salt- and freshwater, which can be converted indirectly into electricity by exploiting the entropy of the two waters (i.e. the propensity of the salinity of mixed liquids to equalise). An electrical current can be produce by mixing the two liquids via a semipermeable membrane. The resource potential for salinity gradient technology is estimated to be 2.6 TW (terawatt). For more information regarding salinity gradient technology, see Jones, A. T., and Finley, W. (2003) Recent Developments in Salinity Gradient Power. In: *Oceans 2003 Proceedings*, 4, pp. 2284–2287.

10 Global Wind Energy Council (2013) *Global Wind Report: Annual Market Update 2013.* Available at: www.gwec.net/wp-content/uploads/2014/04/GWEC-Global-Wind-Report_9-April-2014.pdf

11 REN21 (2017) *Renewables 2017 Global Status Report*. Paris. Available at: www.ren21.net/gsr-2017/

12 Wave energy is essentially a form of wind energy as waves are a function of the energy transfer caused as the wind passes over the oceans. The longer the distance the wind has travelled, the 'fetch', the more powerful are the resultant waves.

13 Salter, S. (2008) Looking Back. In: Cruz, J. (ed.) *Ocean Wave Energy: Current Status and Future Perspectives.* Bristol: Springer, pp. 7–39.

14 Winskel, M., et al. (2006) Energy Policy and Institutional Context: Marine Energy Innovation Systems. *Science and Public Policy*, 33(5), pp. 365–376.

15 Siemens, Alstom and DNCS have been involved in the acquisition of two tidal turbine companies. See Pirolli, B. (2012) French Dive into Tidal Energy as Nuclear Plants Bid Adieu. *ZDNet*. Available at: www.smartplanet.com/blog/global-observer/french-dive-into-tidal-energy-as-nuclear-plants-bid-adieu/7395

16 US Navy Funds Wave Energy Test Site in Hawaii (2014) *Naval Today*. Available at: http://navaltoday.com/2014/07/22/us-navy-funds-wave-energy-test-site-in-hawaii/

17 First-of-Its-Kind Wave Energy Farm for Largest Naval Base in Australia (2015) *Clean Technica*. Available at: https://cleantechnica.com/2015/03/15/first-kind-wave-energy-farm-largest-naval-base-australia/

18 Carrel, S. (2016) World First for Shetlands in Tidal Power Breakthrough. *The Guardian*. Available at: www.theguardian.com/environment/2016/aug/29/world-first-for-shetlands-in-tidal-power-breakthrough

19 World's First Large-Scale Tidal Energy Farm Launches in Scotland (2016) *The Guardian*. Available at: www.theguardian.com/uk-news/2016/sep/12/worlds-first-large-scale-tidal-energy-farm-launches-scotland

20 Margheritini, L., Hansen, A. M., and Frigaard, P. (2012) A Method for EIA Scoping of Wave Energy Converters – Based on Classification of the Used Technology. *Environmental Impact Assessment Review*, 32(1), pp. 33–44.

21 United States Congress House Subcommittee on Oceanography and Committee on Merchant Marine and Fisheries House of Representatives (1978) *Hearings on Renewable Ocean Energy Resources, June 7, 23, 1978*. House Document Serial No. 95-45 at 131.

22 Steinberg, P. E. (2001) *The Social Construction of the Ocean.* Cambridge: Cambridge University Press.

23 Charter, R. (2007) Life on the Edge: The Industrialization of Our Oceans. In: *Proceedings of Coastal Zone 07*, Portland, Oregon, 22–26 July 2007.

24 Osherenko, G. (2006) New Discourses on Ocean Governance: Understanding Property Rights and the Public Trust. *Journal of Environmentatl Law & Litigation*, 21(2), pp. 317–381.

25 Wright, G. (2015) Marine Governance in an Industrialised Ocean: A Case Study of the Emerging Marine Renewable Energy Industry. *Marine Policy*, 52, pp. 77–84.
26 Kerr, S., et al. (2014) Establishing an Agenda for Social Studies Research in Marine Renewable Energy. *Energy Policy*, 67, pp. 694–702; Wright, G., et al. (2016) Establishing a Legal Research Agenda for Ocean Energy. *Marine Policy*, 63, pp. 126–134.
27 Kerr, S. et al. (2014) ibid.
28 Knight, H. (1976) Legal, Political, and Environmental Aspects of Ocean Thermal Energy Conversion: A Report on an ASIL/ERDA Study. In: Kohl, J. (ed.) *Energy from the Oceans: Fact or Fantasy; Conference Proceedings*. Raleigh: The Center for Marine and Coastal Studies, North Carolina State University, pp. 42–46; Knight, H. G., Nyhart, J. D., and Stein, R. E. (eds.) (1977) *Ocean Thermal Energy Conversion: Legal, Political and Institutional Aspects*. Lexington, MA: Lexington Books; Alexander, L. M. (1978) Book Review: Ocean Thermal Energy Conversion: Legal, Political and Institutional Aspects by H. Gary Knight; J. D. Nyhart; Robert E. Stein. *The American Journal of International Law*, 72(4), pp. 934–935; Nanda, V. P. (1979) *Selected Legal and Institutional Issues Related to Ocean Thermal Energy Conversion (OTEC) Development*. Golden, CO: Solar Energy Research Institute; Krueger, R. and Yarema, G. (1981) New Institutions for New Technology: The Case of Ocean Thermal Energy Conversion. *Southern California Law Review*, 54, pp. 767–819.
29 Knight, H. (1976) n. 28.
30 McLachlan, C. (2009) 'You Don't Do a Chemistry Experiment in Your Best China': Symbolic Interpretations of Place and Technology in a Wave Energy Case. *Energy Policy*, 37(12), pp. 5342–5350; Bailey, I., West, J., and Whitehead, I. (2011) Out of Sight but Not out of Mind? Public Perceptions of Wave Energy. *Journal of Environmental Policy & Planning*, 13(2), pp. 139–157; Simas, T., et al. (2012) Understanding the Role of Stakeholders in the Wave Energy Consenting Process : Engagement and Sensitivities. In: *4th International Conference on Ocean Energy*, Dublin.
31 Alexander, K. A., Wilding, T. A., and Heymans, J. J. (2013) Attitudes of Scottish Fishers Towards Marine Renewable Energy. *Marine Policy*, 37, pp. 239–244.
32 de Groot, J., et al. (2014) Investigating the Co-existence of Fisheries and Offshore Renewable Energy in the UK: Identification of a Mitigation Agenda for Fishing Effort Displacement. *Ocean & Coastal Management*, 102, pp. 7–18.
33 Stallard, T., et al. (2009) Economic Assessment of Marine Energy Schemes. In: *Proceedings of the 8th European Wave and Tidal Energy Conference*, pp. 1118–1127; Hayward, J., et al. (2012) Economic Modelling of the Potential of Wave Energy. *Renewable Energy*, 48, pp. 238–250.
34 See the International Network for Social Studies of Marine Energy (ISSMER) website. Available at: www.issmer-network.org/
35 Kerr, S., et al. (2014) n. 26.
36 The ISSMER Agenda refers to broad marine governance issues, including "policy environment, governance, patterns of resource use, conservation values, and distribution of ownership rights", though legal issues in the Agenda are ultimately confined to "dealing with uncertainty" and "planning processes".
37 Castelos, M. A. (2014) Marine Renewable Energies: Opportunities, Law, and Management. *Ocean Development & International Law*, 45(2), pp. 221–237; Mcdonald, S., and Vanderzwaag, D. L. (2015) Renewable Ocean Energy and the International Law and Policy Seascape: Global Currents, Regional Surges. *Ocean Yearbook*, 29(1), pp. 299–326.
38 O'Hagan, A. M. (2012) A Review of International Consenting Regimes for Marine Renewables: Are We Moving Towards Better Practice? In: *4th International Conference on Ocean Energy*, Dublin; Muñoz Arjona, E., et al. (2012) Navigating the Wave Energy Consenting Procedure: Sharing Knowledge and Implementation of Regulatory

Measures. In: *SOWFIA Project – Third Workshop*, 16 October, Dublin; Wright, G. (2014) Regulating Marine Renewable Energy Development: A Preliminary Assessment of UK Permitting Processes. *Underwater Technology: The International Journal of the Society for Underwater*, 32(1), pp. 1–12; Simas, T., et al. (2015) Review of Consenting Processes for Ocean Energy in Selected European Union Member States. *International Journal of Marine Energy*, 9, pp. 41–59.

39   Wright, G. (2014) Strengthening the Role of Science in Marine Governance Through Environmental Impact Assessment: A Case Study of the Marine Renewable Energy Industry. *Ocean & Coastal Management*, 99, pp. 23–30.

40   O'Hagan, A. M. (2012) *Marine Spatial Planning (MSP) in the European Union and Its Application to Marine Renewable Energy*. International Energy Agency Ocean Energy Systems Implementing Agreement website. Available at: www.ocean-energy-systems. org/ocean_energy/in_depth_articles/msp_in_the_european_union/; Azzellino, A., et al. (2013) Optimal Siting of Offshore Wind-power Combined with Wave Energy Through a Marine Spatial Planning Approach. *International Journal of Marine Energy*, 3–4, pp. e11–e25; Davies, I. M., Watret, R., and Gubbins, M. (2014) Spatial Planning for Sustainable Marine Renewable Energy Developments in Scotland. *Ocean & Coastal Management*, 99, pp. 72–81; Kerr, S., Johnson, K., and Side, J. C. (2014) Planning at the Edge: Integrating Across the Land Sea Divide. *Marine Policy*, 47, pp. 118–125.

41   Wright, G., et al. (2016) n. 26.

Part I

Risk and economics

Chapter 2

Risk and ocean energy

Simon Jude, Andrew B. Gill, Craig Mauelshagen and Edward Willsteed

1 Introduction

Ambitious plans for the large-scale deployment of wave and tidal energy are underway to meet carbon reduction targets, with up to 240 GW (gigawatts) of marine (188 GW wave and 52 GW tidal) energy generation capacity forecast by 2050.[1] Test sites, including the European Marine Energy Centre (EMEC), have been established, seabed leases have been awarded, and the deployment of the first, small-scale, commercial array of devices is underway at the Meygen development in Scotland. Plans for significant ocean energy (OE) developments, including up to 1.6 GW of wave and tidal energy generation planned for the Pentland Firth and Orkney Waters (PFOW) in Scotland, are emerging, with test sites being established around the world.[2]

In parallel, the industry's rapid upscaling is driving the development of numerous wave and tidal energy technologies by companies ranging from small embryonic small and medium-sized enterprises (SMEs) to multinational engineering companies and commercial site developers. This includes innovation in devices, materials, components, and enabling technologies in areas including installation and operations and maintenance.[3] However, the industry is immature, with technology testing and demonstration, using scale devices and pilot arrays, still ongoing to demonstrate robust and reliable technology.

The race to develop the OE industry, including technology and developments, and a potential market worth an estimated £76 billion,[4] has resulted in the increased need to be aware of potential hazards and risks. In this chapter, we discuss such challenges, drawing on our experience and insights gained from numerous research projects in the United Kingdom. These include investigating environmental risks and uncertainties associated with OE, the development of risk and decision-making frameworks to support marine regulators and licensing processes, and our practical experience of the marine renewables industry, including the environmental impact assessment (EIA) process.

2 Risk challenges facing ocean energy

A key challenge facing OE projects is that they involve deploying complex engineered devices and structures in the hostile and unpredictable marine environment, thus making them vulnerable to impacts from the environment (in risk terminology, these represent hazards for the OE devices). However, numerous other factors potentially affecting the hazards and risks associated with OE development also exist, including those related to energy policy, political, society and law. For example, evidence from the author's work with industry and regulatory representatives, as part of the UK Natural Environment Research Council (NERC) and Department for Environment, Food and Rural Affairs (Defra) NERC/Defra RESPONSE project NE/J004251/1, which investigated the risk governance issues surrounding marine mammal interactions, has identified a broader range of potential OE project risks. These span issues including financial, political, regulatory (e.g. consenting delays), as well as those related to monitoring requirements and stakeholder opposition.[5] Concerns regarding potential interactions between OE devices and the environment are factors in many of these risks, and it is apparent that there are potential effects on the devices and interactions with the environment, and vice versa, as well as secondary effects.

Hazards and risks can be complex in nature, with differing spatio-temporal characteristics that can manifest themselves throughout the project life cycle, from pre-development survey and monitoring through installation, operations and maintenance, and decommissioning. Despite OE being a young industry, experience to date has already highlighted that such issues can impact the commercial viability of OE developments. This is a concern because the more mature offshore windfarm (OWF) sector has suffered from unforeseen operating expenses (OPEX), which have resulted in significant project delays and insurance claims, with financiers viewing construction risks as unattractive.[6] With installation and operations and maintenance (O&M) representing significant project costs, it is apparent that the fledgling wave and tidal sectors will need to adopt a greater consideration of project risks, including environmental risks, for example, and the development of organizational capacity across the OE sector to improve efficiencies and reduce costs.[7]

Whilst the harsh marine environment clearly poses challenges to OE deployment including device survivability, concerns regarding potential interactions between OE devices, developments and the environment have emerged. Of particular concern is that wide-ranging stressors – anthropogenic activities or consequences of anthropogenic action impacting receiving organisms (e.g. marine mammals and fish) or natural processes (e.g. sediment dynamics) – associated with OE developments have been identified (Table 2.1).[8] Distinguishing how such receptors are affected by OE devices and developments and whether such effects have an actual impact on the receptor or on other receptors, for example through indirect mechanisms, is of critical importance for the OE industry and its regulators.[9] This is further complicated by the need to consider potential positive and negative impacts associated with both individual and multiple/cumulative impacts, as well as whether such impacts are significant at different phases of the project life cycle – all of which are affected by numerous factors, including differing OE technologies and their installation and operation, and have to be interpreted in the context of the current baseline environmental conditions.

Table 2.1 The stressor and receptor links between OE and the ecological components of the marine environment[11]

Level 1 Ocean energy	Level 2 Environmental stressors	Level 3 Environmental receptors	Level 4 Environmental effect	Level 5 Environmental impact	Level 6 Cumulative impact
Wind Wave Tidal OTEC	Physical presence Dynamic effects Energy removal Chemical Acoustic Electromagnetic	Physical environment Pelagic habitat Benthic habitat and species Fish and fisheries Marine birds Marine mammals Ecosystem and food chain	Single/short-term Single/long-term Multiple/short-term Multiple/long-term	Population change Community change Biotic process alteration Physical structure/process alteration	Spatial Temporal Other Human activities

A major challenge is that establishing some useful threshold is seen as critical, particularly by decision makers, but is problematic because of the emergent nature of OE and the associated lack of operational experience and scientific understanding. Hence thresholds inevitably have a level of error associated with them, and so their definition needs to be clear in order to ensure that valid conclusions are drawn from them.[10] Notable challenges exist with regard to the current understanding of the stressor–receptor relationships associated with OE, with some appearing to be well researched whilst others have very little research evidence and remain uncertain. This poses difficulties for both the industry and regulators because the greater the knowledge on stressors, the greater the confidence will be in determining the type of effects on vulnerable receptors. However, this is highly dependent on knowledge of which organisms or processes are vulnerable to change. Worryingly, our research and experience suggest that current research commonly focuses on effects on receptors rather than impacts, with actual impacts often assumed but not demonstrated. This situation is proving highly problematic, in some cases resulting in the OE industry and its regulators placing an undue focus on particular receptors. For example, some potential hazards to particular receptors are considered as likely, even though the evidence does not actually exist. This ultimately encourages risk assessments to be narrow in scope, with uncertainties regarding the true impacts of risks to the whole system remaining. This is problematic because the quality and availability of data, information and knowledge are a critical component of evidence-based risk assessments. As such, future research across all stressors and receptors is vital to address such evidence gaps.

Our research investigating the handling of risk and uncertainty in the OE industry and the risk governance challenges in the Defra/NERC understanding of how OE device operations influence fine-scale habitat use and behaviour of marine vertebrates (RESPONSE) have identified wide-ranging risk assessment, management and governance challenges facing the OE industry, spanning political, stakeholder, investment, project, technology development and regulatory risk.[12] Both our research, which involved detailed reviews of current policy and practice, accompanied by detailed surveys and interviews with representatives from the OE industry and its regulators, and work by the Offshore Renewables Joint Industry Programme[13] have identified poor risk literacy and understanding of fundamental risk concepts, as well as a lack of expertise across the sector as issues constraining expansion. Evidence and insights from our research suggest that the awareness and understanding of risk are still developing across the OE industry. Risk assessment, management and governance practices across the industry are currently immature, with many organizations being small start-ups or SMEs, and are reporting insufficient resources and in-house expertise as limiting their ability to effectively manage risk. This point was emphasised by one survey respondent in the risk and uncertainty project, who noted, "As a small team, the formal risk management procedures can often be overlooked due to other key tasks that may take priority at the time [sic]". At a fundamental level, our research reviewing existing information and literature relating to OE risk and uncertainty and risk assessment practices has identified that variability in what is termed risk has become apparent. Commonly, either the risk was not properly understood or environmental impact was considered as a proxy for risk.

With regard to environmental risks, evidence from an OE industry survey conducted as part of the risk and uncertainty project suggests that the OE industry views the potential environmental risks associated with OE technologies as being low except for those related to consenting. Indeed, as illustrated by Table 2.2, commercial risks arising from current and future consenting processes represent the greatest risk facing the OE sector because they constrain testing and deployment, which are vital if ongoing financial investment is to be secured.[14] The consenting process emphasis is reflected in the environmental impacts research needs expressed by those organizations that engaged in the research, which focus on demonstrating the low environmental impact of renewable energy devices and on reducing the burden of monitoring and EIA for consenting. However, the apparently low level of concern regarding the potential environmental impacts associated with marine energy devices and developments, as well as the risk that the environment poses to them, is of concern. It suggests a possible lack of awareness of the risks that environmental processes (e.g. extreme events, sedimentary processes, mammal collisions) may pose to devices and developments (e.g. physical or reputational damage), despite the experiences from OWF highlighting the vulnerability of offshore infrastructure to environmental hazards and interactions. Concerns regarding wider social and reputational risks – many of which are also linked to environmental interactions – are only beginning to emerge. Rather, the main perceived risks are those impacting commercial and technological development, reflecting the entrepreneurial nature of an emerging and dynamic industry. Thus, the findings suggest that environmental risks are yet to be viewed as representing a potential reputational risk for the sector at present,

Table 2.2 The main risks facing OE organizations as identified by Gill et al.[15]

Survey question	Responses from survey participants	Example quote
What are the main risks facing your organization?	Future regulation that will impact commercial environment	"[U]ncertainties in the renewable obligation certificate post 2017 is [sic] increasing risk to projects (especially arrays which are likely to begin operation in 2017), as investors for these projects require a level of certainty on their return which currently cannot accurately achieved".
	The current consenting processes	"[I]mmaturity of the consenting process. It is an evolving process that is hampered by a lack of regulatory capacity and experience (i.e. Marine Scotland need more personnel)".
	Lack of EIA and environmental monitoring guidelines	"[N]o international standard for marine energy environmental assessment, thus need to prove ourselves with demo projects and licensing in each and every region".
	Unforeseen regulatory constraints due to marine protected areas	"Inability to develop commercially viable sites due to Environmental Designations (SPA's and SAC's) – From a consenting perspective, there is a correlation between the numbers of environmental designations to good wave energy sites".
	Risk to gaining consent due to uncertainty regarding impact on marine life	"[M]arine mammal collision or disturbance effects that are currently unknown and in turn result in long consenting processes which delay a project's manufacture and installation".
	Difficulty securing funding	"[L]lack of funds for project developers".
	Employee health and safety	"Health, Safety and Environment risks to personnel".
	Future consenting and regulatory environment	"The ultimate threat to the organisation is that the technology being developed cannot be installed in the water due to consents not being secured or due to inappropriate conditions placed upon developers as part of the consents or licenses that are permitted".

although this may change as the sector matures from technology innovation and testing to full-scale deployment of devices and arrays.

With the OE industry in its infancy, numerous challenges associated with uncertainties are manifesting themselves. Sources of uncertainty of concern to the OE industry identified by our research include gaining project funding, environmental impact at the commercial scale, and project management and scientific uncertainties including 'unknown unknowns'. However, concerns regarding uncertainties are currently dominated by those surrounding potential impacts on the marine environment and the associated consenting and regulatory environment because consenting represents a key barrier to commercial development. As already noted, the lack of scientific knowledge and evidence of effects and impacts associated with environmental interactions is making it difficult to distinguish between perceived and actual hazards/risks, potentially resulting in perceived hazards/risks being treated as actual. Significantly, knowledge gaps and uncertainties, for example potential impacts on marine mammals, appear to be exacerbated by the lack of clear guidance from regulators and the inexperience across the sector. Such uncertainties are posing numerous challenges, resulting in requirements for technically challenging, costly and pioneering environmental monitoring and environmental impact assessments (EIAs) to demonstrate a lack of negative effects and impacts, which can delay project consenting. Furthermore, the rapid pace and scale of development, along with the accompanying uncertain vulnerabilities, risks and opportunities, are outpacing regulators, who have been viewed as falling back to the precautionary principle and viewed by some in the OE industry as adopting disproportionate monitoring and mitigation requirements.

Evidence from the RESPONSE and risk and uncertainty research projects[16] has also highlighted that the OE sector lacks a cohesive strategy for assessing and managing environmental risk. Whilst aspects of this sector are regulated, guidance is disjointed and unbalanced, and, as previously noted, high levels of uncertainty regarding environmental risks exist. Furthermore, risk assessments are sparse and generally supplemented or substituted with the outputs of the environmental impact assessment (EIA) process. EIA considers an activity's effects on the environment, the spatial and temporal variability of those effects, the cumulative nature of the effects and the sensitivity of the environment to the threat. However, EIA does not provide an understanding of the risks arising from a particular activity. Whilst guidance documents for conducting EIAs for OE development exist, few suggest an assessment of risk when considering strategies for mitigating impacts. Of the EIA guidance that does provide some form of assessment of likelihood of adverse effects, it is often insufficient to inform a complete assessment of risk. As such, EIA and risk management tools, training and guidance are required to support and upskill both the industry and regulators. Increased availability of EIA/risk tools and improved risk literacy is particularly required to address current consenting challenges relating to the lack of regulator clarity on risk management/EIA processes and to industry expertise in implementing and interpreting those processes.

The present situation likely reflects the current engineering and commercial focus of the industry, centring on proving the technology, which requires consents. This can result in risk management being overlooked especially in small organizations. Raising awareness of environmental risks and especially the understanding of the breadth of environmental risks potentially affecting the industry, not just those associated with short-term consenting issues, is required. Analogues from other sectors exist, but it appears that the OE industry can still learn from the risk challenges experienced by offshore wind farm (OWF) development. However, whilst many OE organizations are best characterised as being risk seeking and entrepreneurial, it is foreseen that these challenges will gradually be addressed as the sector matures, with technologies converging and a smaller number of commercial device manufacturers emerging. In parallel, as the sector and organizational maturity progresses, they may become less risk seeking, with formal risk management processes gaining importance both within individual organizations and sectorally. It may also address the issues of limited resources and expertise currently hampering risk management capability. Furthermore, the involvement of large, established power and engineering organizations may result in the involvement of risk assessment and management specialists in OE. It will be interesting to observe such changes alongside the emergence of new forms of risk as the industry evolves and shifts from device testing to the deployment of devices and ultimately large arrays.

3 The need for transparent risk and decision-making frameworks

It is apparent that to facilitate the management of OE environmental risks and uncertainties, evidence-based approaches to both marine licensing and enforcement processes are necessary. This requires the application of risk-based approaches, allowing for uncertainty and recognising the need to use sound science; the need to be sensitive to any potential impacts on sites of particular significance; and the mitigation of negative impacts throughout the development's life cycle. This includes the potential use of licensing conditions or alternative sites or designs, for example using adaptive management strategies, to mitigate the effects, and engagement and discussions between stakeholders and license applicants to resolve any issues.

Risk frameworks are therefore vitally important in guiding decision makers through the risk assessment process, whilst risk assessments comprised of a logical sequence of structured, repeatable and transparent steps enable decision makers to fully comprehend the extent of risk posed by an activity. Significantly, risk assessments enable the identification of those issues posing the greatest risk or threat, allowing risks to be prioritised and mitigated, as opposed to attempting to manage all issues regardless of their likelihood of occurrence. This is important because at present, most EIA guidance provides a structured approach for identifying, investigating and assessing impacts. Thus, they are not unlike risk frameworks, which provide similar clarity and rigour to investigating an issue.

However, in terms of OE risk assessment, EIAs suffer a number of critical limitations. Firstly, they provide only a partial assessment of impact, ignoring consideration of impacts to human health and well-being or the economy. Secondly, they do not provide an explicit statement of likelihood of occurrence, thus limiting the capacity of the EIA to inform risk comparison. Thirdly, they do not provide a common basis (i.e. risk) for conducting prioritisation or options appraisal. Fourthly, they inadequately address the needs of the decision makers who require investigation of the so-what? issue; an activity may have an impact, but how likely is that impact to be realised?

Clearly, transparent decision making, supported by risk-based processes, is vital if the OE sector is to address risk and uncertainty. Developing integrated risk frameworks for the OE sector, which could be employed uniformly across government, industry and academia, would be highly beneficial, providing a logical structure, enabling organizations to assess a variety of different risks and issues from a common perspective. Significantly, such frameworks would ensure the consistent use of terminology, method and assessment criteria. The ability to assess risk in this way is important for providing consistent and relevant risk information to decision makers. Vitally, the adoption of risk frameworks would facilitate comparisons between different risks in a fair and transparent manner and provide an integrated approach to understanding the full suite of risks posed by OE activities. Critically, they would ensure that the same information is available for developers, local authorities and regulators, thus enabling risk-informed and transparent decision processes across the sector. Integrated frameworks would be a first step towards improving decisions about risk and benefit.

Figure 2.1 outlines a proposed integrated framework, based on those developed in other more risk mature sectors, that we believe should be considered as a starting point from which to develop a balanced and integrated approach to dealing with environmental risk and uncertainty. The framework would provide a structured means for assessing and comparing the variety of risk issues affecting the OE sector (e.g. environmental, economic, social). It would extend the process boundaries to consider both direct and indirect impacts across multiple levels. This needs to be achieved in a fair and transparent manner to provide consistency and comprehensiveness to environmental risk and uncertainty assessment. Such an integrated risk framework, developed for the OE sector, could be employed uniformly across government, industry and academia. Significantly, it would provide the structure for which evidence may be incorporated to address and mitigate environmental risk and uncertainty. The framework would also guide research needs, facilitating the identification and targeting of resources to areas where further research and evidence are necessary. Furthermore, it would also indicate appropriate tools at the appropriate scales to reduce uncertainty. From the industrial perspective, an integrated framework would indicate the focal points requiring management in terms of their own assessment of environmental risk.

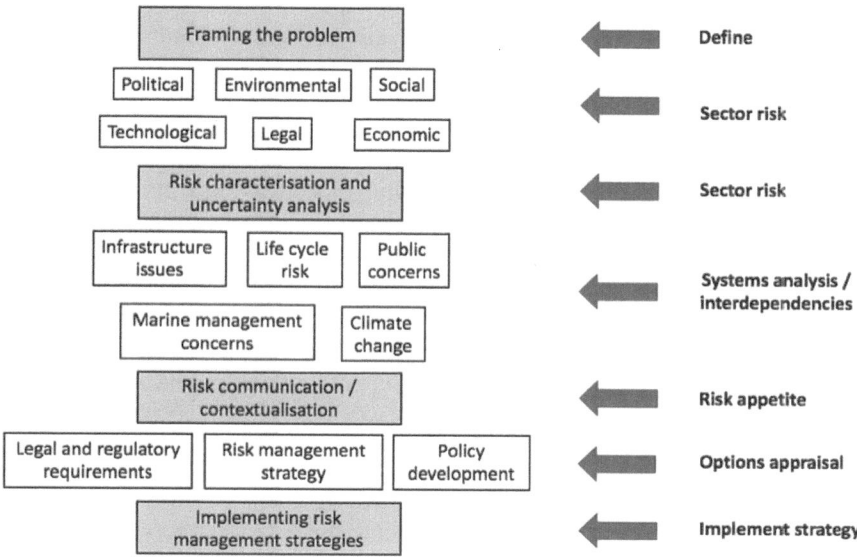

Figure 2.1 Adapted framework for integrated risk assessment and analysis based on Ram (2009)[17]

The outlined framework presented in Figure 2.1 links sector-specific risk assessments with larger, systems-oriented analysis, which then feeds into a risk comparison and risk management strategy development. Importantly, the framework requires that no risk assessment be completed in isolation. Rather, the risks posed by an OE activity should be considered in the broader operational context of the organization, no matter their magnitude, to ensure that risk management strategies developed are both appropriate and cost-effective. The framework is not purely linear and includes a series of feedback loops that enable stakeholders and decision makers the opportunity to reassess decisions in the event that further information is made available (i.e. an adaptive framework). In this respect, the framework can become a living document that requires continual update and evaluation.

4 Methods to communicate risk and uncertainty

Our research has highlighted that, in addition to decision-making frameworks, the development of robust, yet simple and understandable methods for characterising, communicating and comparing uncertain OE risks, accompanied with official guidance, is required alongside the building of capacity and confidence in uncertainty methods. Here the OE industry can potentially build on the existing approaches and experiences in other fields, such as climate change, where approaches for explicitly considering, characterising and communicating uncertain risks and weight of evidence are applied.[18] Furthermore, we strongly believe that innovative approaches for communicating uncertainties and visualising the marine environment are urgently required. Such tools are of importance if the potential risks and opportunities relating to the future growth in OE are to be discussed in the participatory decision-making processes that will be required for the envisaged expansion of the industry to be realised.

To address this challenge, we have undertaken a preliminary exploration of interactive methods for communicating risks and uncertainties using tablet devices as a means of communicating the results from risk assessments. The application of interactive graphics and tablets offers new approaches for communicating the results of risk assessments, and, based on the work of Prpich et al., we have used them to present the findings from a hypothetical OE risk assessment and characterise the nature of environmental harms.[19] By employing simple graphical techniques, complex risks and their accompanying uncertainties and characteristics (e.g. temporality) may be communicated in a simple and comprehensible manner. Our initial work has already illustrated the potential of such technology, which we believe may prove extremely valuable in a range of situations, from discussions among scientists, policymakers and developers to engagement with local communities. However, for such tools to be used effectively by decision makers, it is crucial to build capacity and confidence in uncertainty methods. Currently, uncertainty acts as a barrier to progress/decisions in many sectors. We believe a combination of training, along with the use of robust methods with simple outputs, will raise the utility of uncertainty assessments. For example, numerous methods are available to communicate confidence – many of them explicitly highlight information gaps and would prove beneficial in the OE sector. In parallel, conducting sensitivity analysis would help to answer the so-what? question relating to uncertainty, allowing decision makers to determine whether the level of uncertainty is important to their decision.

5   Developing organizational capacity and community of practice

It is highly apparent that there is a clear and potentially urgent need to raise awareness of risks across the sector and that a suite of EIA and risk management tools, training and guidance is required to support and upskill both the industry and regulators. Only once OE organizations incorporate all types of risk into their activities can they work to achieve a mature risk management capability and begin to benefit from and generate value from this activity. This is vitally important because good risk management offers a competitive advantage by increasing resilience, improving corporate social responsibility initiatives and providing a degree of 'reputational protection' – all key factors in gaining investor confidence. Clearly the development of a community of practice – mostly self-organising communities with shared goals and resources who regularly engage in a common practice – is required. Community members continuously learn together through mutual engagement, requiring the open engagement of regulators and industry with a range of experts to assess, critique and update current frameworks and the risk assessments. However, our experience suggests that to achieve this, there is a critical need to specifically address the linked issues of engagement and information sharing. Worryingly, however, the development of a community of practice is being hampered by the significant commercial pressures and sensitivities associated with the development of the fledgling OE industry. Significantly, our research suggests that such issues, together with an element of commercial secrecy, are impacting potential cooperation and shared learning, despite their potential to build risk literacy and thus avoid potentially significant business risks.

We envisage that the development of mature risk management practices across the OE sector could be achieved through the provision of training and workshops, in addition to accompanying transparent risk-based processes and practical guidance documents that provide step-by-step tutorials on how good risk management can be incorporated into existing practices. A programme of intense knowledge exchange is required to raise awareness of the environmental risks associated with the OE industry. A targeted combination of conferences, free online webinars, secondments, guidance documents, e-newsletters, online resources and face-to-face meetings is likely to enhance the level of awareness amongst developers and operators. Importantly, knowledge exchange and evidence of environmental impacts should be communicated and discussed in an open forum among industry, academia and government, with the overall goal of working together to facilitate a sustainable industry.

To facilitate knowledge uptake and address current gaps in knowledge, we recommend the generation of an online 'OE evidence base' service that aims to regularly review new published evidence in the field and that translates this knowledge into short, useable 'evidence snippets' that are uploaded to an online system. Individuals undertaking a risk assessment could then access the database and quickly access existing information related to likelihood, modelling studies, receptor impacts/sensitivity and more. We envisage such a tool would be useful to regulator, industry, and academic users and could include a monthly/quarterly newsletter service alerting its users to new evidence and research. It would also facilitate the identification of key research needs, allowing current gaps in knowledge to be addressed by conducting research in a balanced way, across all stressors and receptors. Based on the experience of the authors, it is important to ensure that the organization tasked with this important knowledge exchange activity has the capacity to develop such tools.

6 The need to embrace complexity

Alongside the need to develop risk frameworks, the means of communicating risk and uncertainty, and developing a community of practice, there is a need for the OE industry to embrace the complexity associated with hazards and risks.

For example, the vulnerability of OE developments to long- and short-term climate and environmental variability to low-frequency high-consequence events, including wildcards, and to cumulative effects is often overlooked (e.g. UK Climate Change Risk Assessment; Adaptation Reporting Power) and requires further exploration, as do potential opportunities (e.g. artificial reefs) and risk/reward trade-offs. Furthermore, vulnerabilities need to be identified across the project life cycle, especially decommissioning, with greater awareness of the dynamism of emergent technologies and their associated resilience/vulnerabilities, including cumulative/interdependent interactions with other marine activities, and their implications for the industry and regulators.[20] This is important given the increasing complexity and systemic risks associated with large-scale OE, which is increasingly seen as critical national infrastructure and are poised to play a role in national energy security. For example, multipurpose platforms, incorporating OE with other activities, such as aquaculture and energy storage,[21] both potentially reduce commercial risks by diversifying device use and increase potential systems complexity, cumulative and interdependent risks and vulnerabilities.

The application of futures research methods would therefore appear to be of potential benefit to the OE industry. In particular, horizon scanning techniques help to support risk assessments and the identification of risk management and governance options by helping to identify high-probability, high-consequence events, also known as unknown unknowns or wildcards,[22] or extreme impact events, which in hindsight could have been predicted and are commonly known as black swans.[23] Horizon scanning techniques are also beneficial because they enable the identification of emerging areas of interest or concern, be they hazards, risks or opportunities.[24] In addition, scenarios – plausible descriptions of the future – are potentially valuable for the OE industry because they enable a range of alternative futures for a whole system, such as the OE industry, to be considered. Importantly, scenarios are underpinned by coherent and consistent assumptions regarding key factors and their drivers. Already used to support marine planning, scenario building would therefore enable the OE industry to explore future risks and opportunities and to develop long-term risk management and governance strategies and policies, for example.[25]

7 Conclusion

In conclusion, whilst it is clear that the fledgling OE industry faces risk analysis, management and governance challenges as it strives to meet ambitious growth targets, many of the issues identified represent common risk challenges associated with the development of emergent technologies (e.g. nanotechnology) and scientific understanding (e.g. climate change). Thus, as we have illustrated, many of these challenges may be addressed through the application of simple tools and techniques commonly applied in other fields. In parallel, there is a need for more detailed consideration of the complex forms of hazards and risks associated with OE, many of which are emergent in nature and poorly understood. Addressing such complex hazards and risks will require the development of more advanced tools and techniques, together with associated risk management and governance responses. Critically, though, there is an urgent need for the OE industry to develop a community of practice, sharing knowledge and expertise on common risk issues, although it is recognised that this represents a significant challenge for such an embryonic industry.

Acknowledgements

We would like to thank all of the organizations and individuals who have supported our research over several years during a period of significant change in the OE sector. We also wish to thank the Natural Environment Research Council (NERC) for funding the numerous research OE risk projects that this chapter builds on. These include the NERC Marine Renewable Energy Knowledge Exchange Programme Review of the current methods and tools to support policy development and decision making in relation to managing environmental risk and uncertainty in the OE sector; NERC/Defra Understanding how OE device operations influence fine-scale habitat use and behaviour of marine vertebrates – RESPONSE Project (NE/J016330/1); VertIBase, supporting evidence-based decision making on marine vertebrate interactions with wave and tidal energy technologies (NE/01765X/1); and Ed Willsteed's NERC Industrial CASE studentship, investigating marine cumulative effects assessment (NE/L009668/1), co-funded by the Centre for Environment, Fisheries and Aquaculture Science (Cefas). The chapter also builds on our experiences gained through numerous Marine Management Organisation–funded risk and futures projects.

Notes

1 RenewableUK (2015) *Wave and Tidal Energy in the UK: Capitalising on Capability: A Report for the Marine Energy Programme Board February 2015*. London: RenewableUK.
2 Carbon Trust (2011) *Accelerating Marine Energy: The Potential for Cost Reduction – Insights from the Carbon Trust Marine Energy Accelerator*. London: Carbon Trust.
3 RenewableUK (2015) n. 1.
4 Carbon Trust (2011) n. 2.
5 Jude, S. (2012) *RESPONSE Project Risk Scoping Workshop Summary*. Unpublished Report, Centre for Environmental Risks and Future, Cranfield University, p. 7.
6 Carbon Trust (2012) *Offshore Wind Cost Reduction Task Force Report*. London: Carbon Trust.
7 Ibid.
8 Boehlert, G. W., and Gill, A. B. (2010) Environmental and Ecological Effects of Ocean Renewable Energy Development – A Current Synthesis. *Oceanography*, 23, pp. 68–81.
9 Ibid.
10 Wilding, T., et al. (2017) Turning off the DRIP ('Data-Rich, Information-Poor') – Rationalising Monitoring with a Focus on Marine Renewable Energy Developments and the Benthos. *Renewable and Sustainable Energy Reviews*, 74(1), pp. 848–859.
11 Boehlert, G. W. and Gill, A. B. (2010) n. 8.

12 Gill, A. B., et al. (2012) *Review of Environmental Risk and Uncertainty for Supporting Policy Development and Decision Making for the Marine Renewable Energy Sector.* Report prepared for NERC Marine Renewable Energy Knowledge Exchange Programme, p. 84; Jude (2012) 5.

13 Offshore Renewables Joint Industry Programme (ORJIP) (2016) *ORJIP Ocean Energy: The Forward Look: An Ocean Energy Environmental Research Strategy for the UK.* Report to The Crown Estate, Marine Scotland and Welsh Government. Issued by Aquatera Ltd and MarineSpace Ltd.

14 Gill, A. B., et al. (2012) n. 12.

15 Ibid.

16 Ibid.

17 Ram, B. (2009) *An Integrated Risk Framework for Gigawatt-scale Deployments of Renewable Energy: The U.S. Wind Energy Case.* Subcontract Report NREL/SR-500-47129. National Renewable Energy Laboratory.

18 U.S. Climate Change Science Program (2009) Best Practice Approaches for Characterizing, Communicating, and Incorporating Scientific Uncertainty in Climate Decision Making. In: Granger Morgan, M., et al. (eds.) *U.S. Climate Change Science Program Synthesis and Assessment Product 5.2*, p. 88; Intergovernmental Panel on Climate Change (IPCC) (2010) *Guidance Note for Lead Authors of the IPCC Fifth Assessment Report on Consistent Treatment of Uncertainties.* IPCC, p. 6; Skinner, D. J. C., et al. (2014) Identifying Uncertainty in Environmental Risk Assessments: The Development of a Novel Typology and Its Implications for Risk Categorization. *Human and Ecological Risk Assessment*, 20(3), pp. 607–640.

19 Prpich, G., et al. (2011) Character of Environmental Harms: Overcoming Implementation Challenges with Policy Makers and Regulators. *Environmental Science & Technology*, 45(23), pp. 9857–9865.

20 Willsteed, E., et al. (2017) Cumulative Environmental Change: Re-thinking the Effects of Marine Renewable Energy Developments. *Science of the Total Environment*, 577, pp. 19–32.

21 Perez-Collazo, C., Greaves, D. and Iglesias, G. (2015) A Review of Combined Wave and Offshore Wind Energy. *Renewable and Sustainable Energy Reviews*, 42, pp. 141–153; Stuiver, M., et al. (2016) The Governance of Multi-Use Platforms at Sea for Energy Production and Aquaculture: Challenges for Policy Makers in European Seas. *Sustainability*, 8(4), p. 333.

22 Petersen, J. L. (2008) *A Vision for 2012: Planning for Extraordinary Change.* Golden, CO: Fulcrum Publishing.

23 Taleb, N. N. (2007) *The Black Swan: The Impact of the Highly Improbable.* New York: Random House.

24 Garnett, K., et al. (2016) Integrating Horizon Scanning and Strategic Risk Prioritisation Using a Weight of Evidence Framework to Inform Policy Decisions. *Science of the Total Environment*, 560–561, pp. 82–91.

25 Finger, M., Allouche, J. and Luis-Manso, P. (2007) *Water and Liberalisation: European Water Scenarios.* London: IWA Publishing; Marine Management Organisation, Centre for Environmental Risks and Futures and Z_punkt (2011) *Scenarios for the Marine Management Organisation 2031: Final Report.* Marine Management Organisation, Centre for Environmental Risks and Futures and Z punkt.

Chapter 3

Government policy, risk and investment timing

Shelley MacDougall

I Introduction

The governance of wave and tidal stream technologies can be designed to protect the environment and other users of the waters. It can also help to de-risk companies' investments in wave and tidal stream energy conversion. The uncertainties and risks inherent in developing ocean energy are varied and substantial. While the costs and many of the risks are borne by private industry, many of the benefits are societal and environmental. As such, it is logical for governments to help innovative companies overcome financial barriers to help meet renewable energy targets, reduce greenhouse gas emissions, and foster innovation, economic development and rural industrial diversification.

While governance can help reduce uncertainty, it can, in itself, be uncertain. The challenges inherent in the governance of the ocean can result in the unpredictability of government policies, regulations and supports for ocean energy development.

This chapter describes the investment decision faced by technology and project developers, the value of and strategic rationale behind delaying investment and the prevalent uncertainties giving developers pause. It concludes by describing the government policies and supports that can help advance the industry.

2 The investment

Wave and tidal energy conversion technologies are still in the early stages of development. To make it to this stage, a great deal of research, development and testing has been done, work often funded by government and supported by university research. In tidal energy conversion (TEC), which is further along than wave (WEC), we are witnessing the first deployments of the full-scale devices and small demonstration or pilot arrays in commercial waters. Other TEC designs are still in the part-scale testing and demonstration stages.

In the case of TEC, deployments of commercial-scale devices and pre-commercial arrays have met with a fair amount of delay over the years. The delays are, in part, manifestations of uncertainty. There are many uncertainties, both exogenous and endogenous. They generally fall into the following categories: technology, supply chain, construction, operator, political/regulatory, market, and environmental.[1] When uncertainty is high, it is difficult to attract investors, and delaying may be more valuable than investing until a time when the uncertainty is less or at least more measurable.

Companies seek to undertake capital investments that will add value. The amount an investment in ocean energy conversion will add to (or lessen) the value of the firm is estimated using the net present value (NPV) method (see Box 3.1).

By far, the largest outlay is the capital expenditure, approximately 65% of the lifetime costs, depending on the design of the technology, most of which

---

Box 3.1   The net present value method[2]

The net present value (NPV) method discounts expected free cash flows at the cost of capital:

$$NPV = \sum_{t=0}^{n} \frac{FCF_t}{(1+i)^t}$$

and

$$FCF = NOPAT - (\Delta NWC + \Delta NFA)$$

where:
$NOPAT$ = net operating profit after tax
NWC = net operating working capital
NFA = net operating fixed assets (capital assets)
t = year of cash flow
n = estimated economic life of the asset
i = weighted average after-tax cost of capital

Free cash flows include the capital expenditure and revenues less operating expenses (after-tax) over the life of the investment.

The discount rate (i) is the cost of capital. The cost of capital is the (after-tax) weighted average of required rates of return by the company's investors. The rates of return required by lenders and equity investors differ so the combination of debt and equity used to finance the project will affect the overall cost of capital. Debt is less expensive than equity because lenders' exposure to risk is less and interest expense is tax deductible.

must be committed upfront. The major elements of the capital expenditure for ocean energy conversion are project development costs; the manufacture/supply of energy converter(s), the gravity base or other method of mooring, electrical components and monitoring equipment; construction of onshore facilities; installation and commissioning costs; periodic overhauls; and decommissioning.[3] Project development costs include resource assessment (mapping, current measurements and modelling); acquisition of permits; project management; and professional services such as accounting, legal advice, and insurance.[4] The permitting process includes an environmental assessment, wildlife surveys, engineering studies, planning and legal activities.[5] The permitting process can take a year or longer and must be completed before construction can begin. The time it takes to go from application to approval is dependent on the complexity of dealing with multiple planning bodies. The project development phase yields a great deal of data that informs the decision to invest further.

Though there are no primary fuel costs, there are ongoing operating costs for the economic life of the energy conversion devices. The operating expenses of an ocean energy project are for monitoring; routine and emergency maintenance; warehousing costs of components; port berthing fees; insurance, legal, banking and accounting fees; amortization, seabed lease fees;[6] and transmission network charges, if applicable.[7]

Revenues will be a function of the rate paid per megawatt of electricity by an offtaker, plus any production-based subsidy legislated by government (e.g. feed-in tariff). The amount of electricity delivered will be a function of the installed capacity, the capacity factor (megawatts produced/nameplate capacity) of the installed devices, and their availability (uptime).

The financial market ascribes value to the firm's free cash flows, which will be reflected in the company's share price. It reflects not only the expected cash flows coming directly from the investment itself (static NPV) but also those potentially coming from the options the investment makes possible. For example, a company with a proprietary tidal energy conversion technology will be of less value, in the view of investors, than one with the same technology plus a seabed lease, secured for eventual development. The lease constitutes an option: the right to deploy devices at a later date. Even if that option is not yet in the money, it has a speculative value (see Box 3.2).

Box 3.2   Strategic net present value

$$Strategic\, NPV = NPV + V_1 + V_2 + \ldots + V_n$$

where:
V = value of real option.

The Nobel Prize–winning option pricing model by Black and Scholes,[8] adapted for use with optional investments in real assets,[9,10] estimates the value of real options in a manner consistent with the equity valuation principles in finance. The model is shown in the Annex.

There are various types of real options, such as timing options (options to defer investment, delay stages); growth options (follow-on investments, e.g. to expand capacity, expand geographically); options to switch inputs or outputs; shutdown/restart options (to temporarily shut down and restart operations); and abandonment options (timing of disinvestment). These add value to a capital investment by offering managerial flexibility to make later decisions that increase project revenues, reduce expenses, or cut losses.

2.1 The value of delay

The option to defer (delay) an investment or a stage of investment has value because, while waiting, conditions change. As well, with the passage of time, some uncertainty may be resolved. Costs may be less and/or more certain, and technology reliability and revenues may be greater and/or more certain. With better cost and revenue estimates, the cost of capital may be less. There are opportunity costs to delaying, though, such as foregone revenues or a pre-emptive entry into the market by a competitor. These all affect the value of the optional project. If conditions do not become sufficiently favourable for investment, the option can be sold or left to expire.

MacDougall[11,12] questioned whether some of the delays observed in tidal energy development were strategic in nature. First, to explore whether there is economic value in delaying a project, the author estimates the net present value (NPV) of a hypothetical investment in a 10-MW (megawatt) array of commercial-scale tidal energy devices in the Bay of Fundy and the value of the option to delay the investment, using publicly available cost data. As shown in Figure 3.1, the NPV of investing is far surpassed by the value of the option to delay ($2,605,409 versus $14,743,203). From a scenario and sensitivity analysis of the values of investing and delaying, MacDougall finds that there is a somewhat persistent, economically rational incentive to delay investment. The greatest sensitivity of the option value is the unpredictability of the cash flows (volatility). This is consistent with option theory: because of the direct relationship and the asymmetrical risk of an option (limited downside risk and unlimited upside potential), conditions of greater volatility favour waiting. This highlights the importance of risk mitigation in advancing ocean energy development because the higher the risk, the higher the value of the option to delay will be, ceteris paribus.

2.2. Strategic rationale for delay

Beyond the economic model, is there a strategic rationale behind the observed delays? A follow-on study into the qualitative aspects of the investment decision was conducted. MacDougall interviewed executives and senior managers of

Net present value calculation

Energy produced MWh/year	33,288
Revenue (using FIT)	15,811,360
Cost of capital	11.0%
PV benefits	113,697,428
PV costs	111,091,969
NPV	2,605,459
Internal rate of return	11.42%

Value of option to delay	$ 14,743,283
Net present value of investing now	2,605,459
Recommendation:	Delay investment

Value of European call option

Input data

Present value of benefits (S)	S	113,697,428
PV of costs at risk free rate ($)	X	142,183,265
Volatility (measured by std. dev.)	σ	40.0%
Risk-free rate	r_f	2.61%
Leakage (value foregone by waiting)	δ	6.7%
Time to expiry (years) – assumed	t	3.00

Sensitivity to volatility, option expiry and discount rate assumptions

σ	Option value	Years to Expiry	Option value	δ	Option value
	14,743,283		14,743,283		14,743,283
0.25	$ 5,705,721	1.0	$ 7,333,237	0%	$ 25,025,259
0.30	8,556,990	1.5	9,953,093	2%	21,467,174
0.35	11,597,376	2.0	11,950,189	4%	18,331,325
0.40	14,743,283	2.5	13,509,767	6%	15,580,726
0.45	17,941,589	3.0	14,743,283	8%	13,179,836
0.50	21,156,561	3.5	15,723,745	10%	11,094,711
0.55	24,362,889	4.0	16,502,106	12%	9,293,133
0.60	27,541,826	4.5	17,115,649	14%	7,744,735

Black–Scholes option pricing model – European call

$$V = Se^{-\delta t}\left[N(d_1)\right] - Xe^{(-r_f)t}\left[N(d_2)\right]$$

where:

$$d_1 = \frac{Ln\left(\frac{S}{X}\right) + \left[r_f - d + \left(\frac{\sigma^2}{2}\right)\right]t}{\sigma t^{0.5}}$$

$$d_2 = d_1 - \sigma t^{0.5}$$

Figure 3.1 Bay of Fundy 10-MW tidal energy array real option analysis of the decision to invest versus wait[13,14]

ten organizations involved in the development of tidal energy in Nova Scotia, UK and/or France regarding their plans to build and install commercial-scale tidal energy devices.[14] While acknowledging the advantages and disadvantages of being the first mover and a fast follower, there was a consensus that it was important for the company to be ready – technically and operationally – before undertaking the investment. Though problems are typically expected when demonstrating new technology, respondents believed negative outcomes would jeopardize future financing, community acceptance and industry progress.

There are a limited number of suitable tidal energy sites and seabed leases around the world. Technology and project developers with the means have staked out claims in multiple sites. Each site has its own set of issues, in terms of the resource, its measurement and characterisation; infrastructure and access to transmission lines; supply chain capacity; space to build out commercial arrays; and government policies, enabling activities, funding and financial supports. Developers appear to be sequencing their investments across the multiple sites, helping to reduce uncertainties by building in stages, logging operating hours and learning, and going next to where the site characteristics and conditions in the corresponding jurisdiction are most suitable. As Titman and Martin note, "[A] firm with financial resources is less likely to be pressured to invest too soon, when the option to wait is valuable."[15]

Developers are acquiring and holding their sites like a portfolio of options, exercising them when the risks and returns fall into line with investors' requirements. "Because firms have limited budgets, have time and resource constraints, while at the same time have requirements for certain overall levels of returns, risk tolerances, and so forth, portfolio optimization takes into account all these to create an optimal portfolio mix".[16]

These observations are consistent with other studies' findings. Bjorgum, Moen and Masden[17] and Lovdal and Neuwman[18] provide examples of marine energy companies using internationalization as a strategy to overcome barriers. As the industry develops globally, barriers, government policies, and supports, which vary by jurisdiction, change over time. Bjorgum et al.[19] conducted multiple case studies of pre-commercial marine renewable energy technology developers in Norway. They found that companies internationalizing in the pre-commercial stage were partnering with companies in other jurisdictions. This was done to not only broaden their technical expertise but also to "exploit international opportunities created by different policy regimes".[20]

These studies suggest that pre-commercial technology and project development companies are mobile. When ready, those with the means may choose to build in other jurisdictions, where the conditions are best suited to achieving their objectives.[21]

2.3. Predominant uncertainties

Being that the uncertainty of cash flows is a key driver in option value, MacDougall inquired further into what respondents believed were the predominant uncertainties they faced in deploying commercial-scale tidal energy conversion devices and what they felt would help move the industry along. There was a fair amount of consensus around six themes, which are summarized.[22]

1 **Technology reliability.** The uncertainty surrounding the ability of the technology to hold up in the harsh marine environment arose as the dominant theme: how well and how long the technology would perform and what the frequency and nature of unscheduled maintenance would be. Deploying and operating devices in tidal streams are integral to reducing this uncertainty.

2 **Site and resource knowledge.** Areas where there is a lack of knowledge of the resource and site characteristics present a large amount of uncertainty. Knowledge of the site and the resource helps developers 'right-size' the turbines, plan their appropriate placement and estimate the energy yield. It also helps developers determine whether their installation procedures and technology designs are suitable and ready for deployment in those particular conditions.

3 **Energetic site and space to build out a commercial array.** An energetic site is important for achieving economies of scale. Related to this is the availability of space to build commercial arrays. This is also needed to generate sufficient economies of scale to bring down the cost of energy to a level that will be competitive with other renewables.

4 **Suitability and predictability of government policy and financial support.** A major area of uncertainty is related to government policy, funding and financial supports. Companies seek sufficient and predictable support that is flexible and patient.

5 **Prospects for offtake agreements.** Even with a feed-in tariff (FIT) in place, the lack of offtake agreements makes raising capital especially difficult.

6 **Supply chain capacity and pricing.** Another major source of uncertainty facing developers is supply chain capacity and the cost of technologies, materials, such as steel, and services. In the relatively unindustrialized and rural coastal communities located near the tidal resources, the supply chain is often insufficiently robust.

These uncertainties are greatly inhibiting developers' ability to finance projects. Government funding and financial support are all the more important to facilitate the building, deployment, operation, maintenance and retrieval of commercial-scale devices in commercial-grade waters so that learning can occur, uncertainties reduced and risks measured.

Also, respondents look to jurisdictions to undertake thorough site assessment and resource measurement, identify sites with potential for commercial build-out in terms of resources and physical space, and provide a path forward for permitting and consenting. They also noted the need for government help to build supply chain capacity and to support its innovation.

3 Government policy and the investment decision

Although the investment in ocean energy conversion is incremental by nature, that is one or several devices at a time, each increment is costly. Building even a pilot array of tidal or wave energy devices constitutes a large, upfront, largely irreversible commitment of capital and a long investment horizon. The cost of energy will be uncompetitive with other sources of energy for quite some time. These, combined with the uncertainties and risks, make the incentive to delay greater and government policy all the more important.

Private-sector financing will remain limited so long as uncertainty is so high. Most of the private-sector capital being invested at this early stage is balance sheet financing by industrial companies with a strategic interest in the industry. While balance sheet financing may be able to meet the costs of research, development, and part-scale and perhaps full-scale demonstration of an ocean energy conversion technology, progressing beyond that point is often beyond the investment capacity of even a large business without changing its own risk profile. The size and investment horizons of these technologies do not fit the venture capital model. Other equity investors, both individual and institutional, will generally seek investments offering a better risk-return trade-off. Bank financing will not likely be available until the technology is commercialized.

Table 3.1 shows the stages of ocean energy conversion technology development, the timing and sources of private sector financing, and the timing and types of suitable financial support by government. It also shows when technology development meets what is often called the technology valley of death,[23] when angel capital and balance sheet financing are no longer available or sufficient, but the uncertainty is still too high for equity markets and banks.

Government policies and programmes help mitigate risk and allow projects to be undertaken and learning to be gained so the industry can develop to one that is self-sustaining. Usually, a combination of funding and financial support mechanisms (e.g. capital grants and feed-in tariffs) are needed to help the industry progress to the stage when private-sector financing can be raised, along with other enabling measures. Enabling measures help remove barriers to development or facilitate progress. They include building publicly funded testing facilities and demonstration sites, constructing infrastructure, streamlining consenting and permitting processes, and collecting and disseminating data collection (e.g. resource measurement and site characterisation).[24,25]

Jurisdictions at the forefront of marine renewable energy research and development have introduced policies and financial supports such as renewable obligations, capital grants, price supports (feed-in tariffs, renewable energy certificates). They have also streamlined consenting and permitting processes and have undertaken site characterisation and resource measurement, community consultation, and infrastructure investment. These can help devices be built, enabling technologies and pilot arrays to be developed and demonstrated. Through these developments, learning can be gained and operating hours logged, which, in turn, reduce uncertainties, allow risks to be measured, and facilitate future private-sector financing.

While government policies can measurably help the business case of an investment in ocean energy development, they are subject to changes in government and budget cuts. Since, at this stage, government support is so integral to the

Table 3.1 Stages of OE development, typical sources of financing, government supports[26]

Stage	R&D[27]	Part-scale demonstration	Full-scale device demonstration	Demonstration array	Pre-commercial array	Commercial arrays
Activities	Generate idea, IP; develop prototype.	Test prototype in simulated environment, refine prototype.	Demonstrate full-scale prototype (device, reaction system, station keeping) in operational environment; prove technical validity in the field.	Demonstrate array-enabling technologies, demonstrate array; refine technologies, processes, materials.	Build small array in commercial marine area; commercial viability not yet achieved, technology costs still high.	Build large, commercially viable arrays.
Sources of private-sector financing	Balance sheet finance, angel investment, corporate venture capital		Balance sheet finance, corporate venture capital 'Technology valley of death'			Bank finance, project finance, public markets, mergers, acquisitions
Government support mechanisms (financial)	R&D grants, government and university research, technology prizes		Renewables quota, price-support (diminishing $/kWh)			
			Demonstration grants, array-enabling technology grants, array-demonstration grants, infrastructure investment, development bank financing		Capital grants, public–private investment fund, tax incentives, domestic and export development bank financing	

viability of ocean energy conversion investments, uncertainty about government support itself deters private-sector investment.

As developers move through the stages, installing demonstration units and demonstration and pre-commercial arrays, government regulations and financial supports need to be clear, sufficient, stable and predictable.[28,29] Carefully defining and communicating the programme's overarching goals will help provide clarity and a line of sight to the time when private-sector financing can be raised. Examples of programme goals are to meet emissions targets, improve energy security, diversify energy supply, support local economic development, diversify rural industry, lower (renewable) energy costs, develop green technology markets, create jobs, and create exportable technologies. Used as a touchstone, the goals also help ensure that the elements of the package (regulations, infrastructure investment, financial supports, etc.) are complimentary and internally consistent.[30]

Though it needs to be predictable, government financial support does not need to be permanent. Nor should it be. As the industry becomes viable on its own, government support should diminish to make way for private-sector investment.[31]

4 Summary

Government policies and regulations can protect marine and coastal areas and the welfare of various stakeholders. They can also help reduce the risks borne by private companies willing to innovate to develop ocean energy resources. The risks are sufficiently high at the early stages and the upfront investment so large that government support is essential for the industry to become established.

There is evidence that there is value in delaying investment in commercial-scale devices under various conditions and in staging investments across multiple seabed leases, and there is a willingness to internationalize in the pre-commercial stage. Developers may well be drawn to jurisdictions where the government policies and supports facilitate the next steps, suggesting that the relative attractiveness of jurisdictions' offerings is important. The real option valuation model is an appropriate tool to help set government ocean energy policy. It can be used by policymakers to design the offering – its package of infrastructure, site knowledge, supply chain, funding and financial supports – that it bundles with its naturally unique tidal resources. The model can be used to estimate the impact of changes to the package's elements on the values of investing and delaying in a manner consistent with the equity market's priorities.[32]

Government policies can also be a source of uncertainty if they are unclear or subject to change. The government programme needs to be clear, sufficient, stable and predictable so as to reduce uncertainty related to policy, financial and market risks. Clearly defined and articulated goals of the government programme can help in this regard. Developers and investors can gauge the breadth, depth and duration of government commitment and have a line of sight to when financing can be raised from private-sector equity investors and commercial banks. As the industry becomes viable on its own, government support can be phased out to make way for private-sector investment.

Annex: Black–Scholes option pricing model

The Nobel Prize–winning option pricing model by Black and Scholes,[33] adapted for use with optional investments in real assets,[34],[35] values real options in a manner consistent with the equity valuation principles in finance. Though it is difficult to capture the exact value of a real option due to the non-tradability of the real asset and the existence of compound options, the model can help the company make capital investment decisions that align with how the market values the firm and its equity.

The Black–Scholes option pricing model for a dividend-paying European stock option is as follows:

$$V = Se^{-\delta t}N(d_1) - Xe^{-r_f t}N(d_2) \tag{3.1}$$

$$d_1 = [\ln(S/X) + (r_f - \delta + \sigma^2/2)t] / \sigma\sqrt{t} \tag{3.2}$$

and

$$d_2 = d_1 - \sigma\sqrt{t} \tag{3.3}$$

where:
V = the current value of the call option.
S = the current price of the underlying share.
X = the exercise or strike price (present value).
r_f = the risk-free rate of interest, continuously compounded.
e = 2.7183.
$N(d)$ = cumulative normal probability density function.
δ = the payout rate or dividend rate on the project.
t = the time left until expiry.
σ = the standard deviation of the stock's rate of return.
$\ln(S/X)$ = the natural logarithm of the ratio of share price to exercise price.

Table 3.2 shows how the option pricing model for a financial asset is applied to a real asset. The variables S and X are similar to the values used in net present value calculations. S, or the present value of expected benefits, is the same in both models. X, or present value of expected costs, differs only by the discount rate used. As well, the period of discounting is different: the European option pricing model assumes the cash will begin to flow upon exercise at the end of the option period, if at all.

Table 3.2 Input variables, relationship to option value, OEC conditions that can affect option[37]

Variable	Definition	Relationship to V	Conditions and changes thereto that can affect the variable
S	Present value of expected benefits, discounted at cost of capital	Direct	Feed-in tariff (amount, duration) Equipment reliability, utilization Energy yield, capacity factor Economic life Grid access
X	Present value of expected costs, discounted at the risk-free rate	Indirect	Reliability affecting maintenance costs Knowledge of resource affecting accuracy of design specifications Supply chain capacity Infrastructure in place Soft investment costs by leader versus follower Water speed, tidal range, distance from shore Capital grant (reduces cost to owner)
σ	Volatility of expected cash flows, measured by standard deviation	Direct	Visibility of government policies, power purchases, future FITs Knowledge of resource, site Technology reliability, operating hours, data collected (knowledge of energy yield, breakdowns, repairs) Performance guarantees Insurance Construction cost uncertainty, delays Interaction with marine environment Likelihood of grid access
δ	Leakage rate (foregone benefit each year by waiting in %)	Indirect	Pace of development Ability to secure rights to develop commercial site Exclusivity of right to develop resource Lease fees Time limit on FIT
t	Time to expiry (years)	Direct	Time limits to develop at site, duration of FIT, capital grants

While the cost of investing (exercising the option) is greater than the value of the inflows ($S < X$), the option is 'out of the money' and it makes economic sense to wait. Once $S > X$, the option is 'in the money' and can be exercised. Theoretically, it is suboptimal to exercise an option until just before expiry because the option value can still increase over the remaining time. However, in practice, managers may find compelling qualitative or hard-to-quantify reasons to exercise before expiry.[36]

Table 3.2 also lists factors that can affect real option value for an investment in ocean energy conversion.

Notes

1 MacDougall, S. (2013) Financing, Government Supports, and Managing Risk. In: MacDougall, S., and Colton, J. (eds.) *Community and Business Toolkit for Tidal Energy Development*. Acadia Tidal Energy Institute, March 2013, pp. 215–241. Available at: http://tidalenergy.acadiau.ca/community-business-toolkit.html
2 Brigham, E. and Daves, P. (2013) *Intermediate Financial Management*, 11th ed. Boston: Cengage Learning.
3 RenewableUK (2011) *Offshore Wind: Forecasts of Future Costs and Benefits*. Available at: www.bvgassociates.com/cases/offshore-wind-forecasts-future-costs-benefits/
4 Entec UK (2006) *Cost Estimation Methodology: The Marine Energy Challenge Approach of Energy Produced by Marine Energy Systems*. Commissioned by Carbon Trust. Available at: www.carbontrust.com/resources/reports/technology/marine-cost-of-energy-methodology/
5 Ibid. n. 3.
6 RenewableUK (2011) n. 3.
7 UK Energy Research Centre (2010) *Investment in Electricity Generation: The Role of Costs, Incentives and Risks*. UK Energy Research Centre. Available at: www.ukerc.ac.uk/Downloads/PDF/06/0706_Investing_in_Power.pdf
8 Black, F., and Scholes, M. (1973) The Pricing of Options and Corporate Liabilities. *Journal of Political Economy*, 81(May–June), pp. 637–659.
9 Dixit, A. K., and Pindyck, R. S. (1994) *Investment Under Uncertainty*. Princeton, NJ: Princeton University Press.
10 Trigeorgis, L. (1996) *Real Options: Managerial Flexibility and Strategy in Resource Allocation*. Cambridge, MA: MIT Press.
11 MacDougall, S. (2014) *Better to Wait? Investigating the Value of Delay of In-stream Tidal Energy Projects*. Abstract and Poster Presentation, International Conference on Ocean Energy, Halifax, Nova Scotia.
12 MacDougall, S. (2015) The Value of Delay in Tidal Energy Development. *Energy Policy*, 87, pp. 438–446.
13 Ibid.
14 MacDougall, S. (2017a) Confronting the Financing Impasse: Risk Management Through Internationally Staged Investments in Tidal Energy Development. *International Journal of Marine Energy*, 18, pp. 78–87.
15 Titman, S., and Martin, J. (2011) *Valuation: The Art and Science of Corporate Investment Decisions*, 2nd ed. Boston: Addison-Wesley, p. 502.
16 Mun, J. (2002) *Real Options Analysis: Tools and Techniques for Valuing Strategic Investments and Decisions*. Hoboken, NJ: Wiley, p. 96.
17 Bjorgum, O., Moen, O., and Masden, T. (2013) New Ventures in an Emerging Industry: Access to and Use of International Resources. *International Journal of Entrepreneurship and Small Business*, 20(2), pp. 233–253.

18 Lovdal, N., and Neumann, F. (2011) Internationalization as a Strategy to Overcome Industry Barriers – An Assessment of the Marine Energy Industry. *Energy Policy*, 39, pp. 1093–1100.

19 Ibid. n. 16.

20 Ibid. p. 250.

21 MacDougall, S. (2017a)

22 MacDougall, S. (2017b) *Strategic Timing of Commercial-Scale Tidal Energy Investment.* 2017 European Wave and Tidal Energy Conference Proceedings.

23 Moore, G. A. (1991) *Crossing the Chasm: Marketing and Selling Disruptive Products to Mainstream Customers.* New York: Harper Business.

24 Low Carbon Innovation Coordination Group (2012) *Technology Innovation Needs: Marine Energy.* Available at: www.carbontrust.com/media/168547/tina-marine-energy-summary-report.pdf

25 Gardner, M., MacDougall, S., Taylor, J., Karsten, R., Johnson, K., Kerr, S., and Fitzharris, J. (2015) *Value Proposition for Tidal Energy Development in Nova Scotia, Atlantic Canada and Canada.* Offshore Energy Research Association, Halifax, NS. Available at: www.oera.ca/wp-content/uploads/2015/04/Value-Proposition-FINAL-REPORT_April-21-2015.pdf

26 MacDougall, S. (2016) *Funding and Financial Supports for Tidal Energy Development in Nova Scotia.* Prepared for the Offshore Energy Research Association, Halifax, Nova Scotia. Available at: www.oera.ca/wp-content/uploads/2016/12/2016-09-09-TE-Funding-Financial-Supports_FINAL.pdf; Bloomberg New Energy Finance (2010) *Crossing the Valley of Death: Solutions to the Next Generation Clean Energy Project Financing Gap.* Available at: www.cleanegroup.org/wp-content/uploads/Crossing-the-Valley-of-Death.pdf; Ghosh, S., and Nanda, R. (2010) *Venture Capital Investment in the Clean Energy Sector.* Harvard Business School Working Paper, pp. 11–20. Available at: www.hbs.edu/faculty/Publication%20Files/11-020.pdf; Kalamova, M., Kaminker, C., and Johnstone, N. (2011) *Sources of Finance, Investment Policies and Plant Entry in Renewable Energy Sector.* OECD Environment Working Papers, No. 37. Available at: /www.oecd-ilibrary.org/environment/sources-of-finance-investment-policies-and-plant-entry-in-the-renewable-energy-sector_5kg7068011hb-en; ORE Catapult (2014) *Financing Solutions for Wave and Tidal Energy.* Available at: www.all-energy.co.uk/__novadocuments/81670?v=635640052099430000

27 ORE Catapult R&D activities are proof of principle, component critical functions, design integration.

28 Burer, M., and Wustenhagen, R. (2009) Which Renewable Energy Policy Is a Venture Capitalist's Best Friend? Empirical Evidence from a Survey of International Clean Tech Investors. *Energy Policy*, 37, pp. 4997–5006.

29 Leete, S., Xi, J., and Wheeler, D. (2013) Investment Barriers and Incentives for Marine Renewable Energy in the UK: An Analysis of Investor Preferences. *Energy Policy*, 60, pp. 866–875.

30 MacDougall, S. (2016).

31 Ibid.

32 MacDougall, S. (2017b) p. 40.

33 Black, F., and Scholes, M. (1973) The Pricing of Options and Corporate Liabilities. *Journal of Political Economy*, 81(May–June), pp. 637–659.

34 Dixit, A. K., and Pindyck, R. S. (1994) *Investment Under Uncertainty.* Princeton, NJ: Princeton University Press.

35 Trigeorgis, L. (1996) *Real Options: Managerial Flexibility and Strategy in Resource Allocation.* Cambridge, MA: MIT Press.

36 Being able to exercise at any time before expiry is a valuable feature of an American option, but its valuation is more complicated. Using the pricing model for a European option is practical and generates a conservative estimate of the option value.

37 MacDougall, S. (2017b) n. 21.

Part II

Marine governance

Chapter 4

Building governance at sea

Kate Johnson

1 Introduction

In setting out to build governance at sea, we do not start exactly with a clean sheet, but it is certainly much cleaner than we would find anywhere on land. If marine space is not subject to full sovereign jurisdiction and not in private ownership, then who or what exactly should be responsible for governance, the ultimate arbiter of the ways in which maritime affairs are carried on? The answer is a growing catalogue of international law and regulation that has tended to be reactive. It responds to visible problems like pollution and tries to anticipate the needs of future industries like seabed mining because such industries are expected to develop. But international law is fragmentary, far from complete, difficult to negotiate and weak in enforcement. It is also complex and confusing as recorded in Sue Boyes' 'Horrendogram' that plots the spider's web of connections among international laws, European laws and national laws insofar as they apply to the maritime environment.[1] An important start in governance is to draw a distinction between this and government. Governance and even good governance can exist without government. Indigenous communities have long exercised forms of control and management over the areas of sea important to them. Many of these traditional forms of governance remain valid to this day, and some will be protected under international law. For example, the maritime rights of the Inuit people in Canada were not extinguished under the colonial regime as were many of their terrestrial rights.[2] In other examples, many groups of fishers exercise governance over the fishery in their areas of interest and enforce it through peer pressure.[3]

However, the overwhelming trend is for increased sovereign rights at sea supported by international law, although sovereignty is limited to varying degrees in every marine geographic zone. The territorial sea (up to 12 nm (nautical miles)) approaches the full sovereign power of government for the coastal state but is diluted by the requirement to allow innocent passage to foreign vessels. The exclusive economic zone (up to 200 nm) grants certain sovereign rights and responsibilities for resources to coastal states but allows high seas (outside 200 nm) rights of international access. It is said that the territorial sea (TS) tends to be contiguous with the land while the exclusive economic zone (EEZ) tends to be contiguous with the high seas (HSs). In areas where sovereign jurisdictions or rights apply, the governments of coastal states take the lead on governance, but they are not necessarily free agents. In the HSs area, the international community has attempted to introduce a United Nations managed regime of governance through the International Seabed Authority but the detailed terms remain in dispute. Everywhere, authorities are constrained by respect for legally protected traditional rights and the various governance options that might be applied. On land, where nearly everything is owned, the government can control the activities of the owners of the property right if it decides to do so. At sea, where virtually nothing is owned, it becomes a matter of cooperation among states, markets and people. International law, supported by treaty and custom, sets the framework, but international law binds only those who agree to be bound. Power relationships are explored in more detail in Chapter 12 of this book, but power in governance has been described as the balance among three factors: (1) the power of the state or equivalent international institutions, (2) the power of the market and (3) the power of the people.[4] This is explored in Section 2 of this chapter.

Governance at sea is a process still under construction. The implications are highly significant as, after millennia of slow evolution, we move more rapidly away from the concept of Freedom of the Seas towards enclosure of the marine commons. The aim of this chapter is to explore the role and place of ocean energy (OE) in these emerging frameworks of maritime governance: first, to identify the actual and desired characteristics of OE and the extent to which they could be shaped by the needs of governance; second, to consider how the most important characteristics of OE might shape or modify governance itself. This chapter sets out to build the story. In Section 2, we look at a theory of governance and the balance of interests between top-down and bottom-up approaches. Section 3 explores the foundations of maritime governance from the early needs of trade, fisheries and security at the time of the Phoenicians (1300 BCE) right through to the middle of the twentieth century. Section 4 records a time, in the late twentieth and early twenty-first centuries, of rapid change and development in maritime governance – spatially defined industries based on fixed platforms becoming established and concerns about pollution and ecosystem services leading to the creation of new environmental management measures. A case study of the governance of wave and tidal energy arrays is made in Section 5 with a focus on the anticipated areas of impact for such an industry. Finally, conclusions and guidance are given in Section 6.

2 A theory of governance

Plato considered the role of the state in the steering of human affairs in his writing of *The Republic*.[5] The word 'governance' itself derives from his use of the Greek verb 'to steer', which in Roman script is *kubernáo*. Subsequent thinkers and philosophers have advanced their ideas towards an understanding of how the basic tenets of governance should be applied and where this 'steering' should come from. Hobbes, in *Leviathan* (1651), argued for the need of a strong central government, in which expert knowledge would guide decision making through state-controlled bureaucracies.[6] People holding such an environmental perspective today were described by Dryzek in 1987 as 'ecological Hobbesians' or 'neo-Hobbesians'.[7] This might be viewed as top-down, or hierarchical, governance. At the opposite end of the scale are those who might be described as neo-institutionalists, promoting a bottom-up, or polycentric, form of governance.[8] This is a sociological rather than an economic approach, with a focus on how institutions can shape the behaviour of people and, in return, be shaped by them. A third approach to the governance of marine resources is to allow markets to find the level at which exploitation should be controlled and managed. Denman and others have argued for markets in the sea, whereby the private sector will judge the value of the resource and manage the resource to protect the investment.[9] This will apply under a *res nullius* (no ownership and no jurisdiction) regime, such as that claimed by some (including the United States) for the high seas. It could be achieved in other jurisdictions through the establishment of tradable private property rights in the resource. Tradable quotas in fish stocks are an example. The buying and selling of seabed leases licensed for aquaculture are another. In summary, we have these three approaches to governance and a perpetual search for a balance among them.

1   **State control or administrative rationality** – a top-down governance process based on expert knowledge guiding state-controlled bureaucracies. An example would be the EU Habitats Directive and the subsequent designation of Natura 2000 areas by national governments. This point of view might be referred to as neo-Hobbesian after Thomas Hobbes' argument for strong central government.[10]
2   **Market forces or economic rationality** – allowing market forces the freedom to find their own level, otherwise known as neo-liberal economics. An OE example is the fixing of boundaries for wave and tidal arrays in the Pentland Firth and Orkney Waters (PFOW) area of Scotland. The government agency responsible for the seabed advertised for developers to tender for rights to occupy the areas of their choice based on a block grid system.[11]
3   **Public interest or deliberative rationality** – an emphasis on civil society and self-organising networks of interested group and individual users, a bottom-up governance process. Examples from the offshore oil industry are the unusually extensive maritime and financial powers awarded to the Shetland and Orkney Island Councils at the time of North Sea oil development in the 1970s.[12]

3 Foundations

Claims to some form of ownership of the seas can be traced at least to the time of the Phoenicians (ca. 1300 BCE). The concept of *mare nostrum* ('our sea') was applied by successive powers in the Mediterranean. While Roman law recognised many common rights to the seas for Roman citizens, clear responsibilities for the protection of coastal waters from piracy and jurisdiction for matters such as salvage were assumed by the state governing the adjacent territory. The areas beyond *mare nostrum* were *mare vastium* (the uncontrolled seas beyond state jurisdiction). *Mare nostrum* was limited by the extent to which a state was able to enforce its jurisdiction, which depended ultimately on naval power, more a matter of force than law. These early concepts, driven largely by the needs of good order in trade and security, found expression in two developing bodies of law. First, Admiralty Law, which deals with matters of marine commerce, navigation, salvaging, shipping, sailors and the transport of passengers and goods.[13] Second, the International Law of the Sea, which deals with rights of navigation, resources, jurisdiction and relations among nations.

The earliest surviving example of a legal declaration for the seas is the Rhodian Maritime Code (*Lex Rhodia*), which was in place around the second and third centuries BCE. There is no written record of the code itself, only much later writings about it. The Rhodian Maritime Code was more concerned with commerce than dominion. It recognised the right of all nations to use the seas freely for the legitimate purposes of transportation of goods and merchandise for trade. At this time, the Aegean Islands of Rhodes, an independent state of the Greek Empire, was in ascendancy within the Greek dominion. By the third century BCE, Rhodes had replaced Athens as the commercial centre of the Greek Empire and was the focus for commerce and trading activity with the Orient, Spain and North Africa. *Lex Rhodia* was not forced on any nation but was recognised voluntarily. The code collected and codified maritime customs and practices and sought to adopt them over much wider areas. It embodied a principle of Freedom of the Seas and the rights of all nations to use the Mediterranean Sea for trade. The code evolved from customary practice and made provisions for liability in cases of injury or damage. Custom had assumed a binding character to protect the interests of trade and commerce. It was not international law in the way it is now understood, but it is an historic example of nations voluntarily binding themselves to rules in shared waters.

The Roman state embraced *mare nostrum* and recognised a responsibility for protection from piracy. Jurisdiction for the purpose of taxation and the right of salvage was assigned to the state. The tradition of Roman law was that seas, rivers and seashores were common to all Roman citizens and could be used by anyone wishing to do so. This classical view is captured by Ovid in his 'Metamorphoses': '*Why do you deny me water? Its use is free to all. Nature has made neither the Sun nor Air nor Waves private property; they are public gifts*". This concept is *res communis* ('of the commons' or 'of common property'), meaning that the seas belong to all and must remain open for all to use. A different concept is that of *res nullius*, which conveys the idea that the seas belong to no one and are thus available for appropriation. Under *res communis*, appropriation infringes the rights of everyone; under *res nullius*, appropriation is open to anyone who has the will and the resources to enforce it, and nobody's rights are infringed. There are many examples of appropriation of sea areas in medieval times. The most ambitious example is the papal agreement to divide the New World discovered by Columbus between the sovereign powers of Spain and Portugal. The Treaty of Tordesillas in 1494 established a line 370 leagues west of the Cape Verde Islands running from the North Pole to the South Pole. Unless otherwise in the possession of a "Christian" power, all lands to the west of this line were deemed Spanish and to the east, Portuguese – no vessels were to voyage to these lands or to fish in their waters without agreement from these sovereign powers. This wave of medieval expansionism depended on the ability to enforce such claims.

Queen Elizabeth, protesting in 1599 to the King of Denmark about the seizure of fishing and merchant vessels in the North Sea, invoked the Law of Nations in favour of the principle of open seas. She invoked what might now be called international law by reference to the Law of Nations. Just a few years later, around 1605, true expression of an international legal principle came with the publication of *Mare Liberum*, a scholarly work by Hugo Grotius of the Netherlands. The Dutch East Indies Trading Company commissioned the work in support of their need for freedom of navigation and free trade. Nonetheless, *Mare Liberum* was a scholarly analysis and an argument in favour of freedom of navigation: "*No part of the sea can be counted in the territory of any people*". In short, the seas were a commons, available for all to use and not available for appropriation. Grotius's arguments were mainly concerned with navigation, but his critics were concerned with fishing and the rights of appropriation as security. The Englishman John Selden (1635) is usually cited as the rebuff to Grotius in his work *Mare Clausum*, which argued that coastal states had the right to appropriate areas of the seas adjacent to their coasts. In all but the most recent developments, the law of the sea can be seen as the marriage of, or compromise between, these two doctrines. On the one hand, the freedom of the seas, on the other the right of a coastal state to appropriate or exercise sovereign rights over an area of the sea adjacent to the coast.

A framework of rights and practice was established through agreement and cooperation among nation states. Treaty and custom supplied the foundation of a body of international law for the oceans and seas that sufficed right through to the twentieth century. However, increasing trade and demand for maritime resources and space exerted pressure for change, which developed slowly through the first half of the twentieth century but accelerated rapidly through the second half of the century and to the present day.

4 Contemporary drivers and directions

By the end of the nineteenth century, many maritime nations had established territorial seas, often extending to 3 nm, which is a distance claimed to have resulted from the range of a shore-based cannon. A nation's territorial jurisdiction was perhaps limited to the extent to which the sea could be defended from the shore. Subsequently a number of states extended claims to 12 miles with a variety of zones included. An attempt was made to resolve the varying extents of territorial seas being claimed when the League of Nations called the Hague Conference in 1930. The final text was not adopted by the Conference owing largely to the failure to reach agreement on a specified width for the territorial sea, mainly as a result of fishing industry concerns. However, it did establish the principle of coastal sovereignty, which has never been seriously challenged. The territorial sea had no basis in treaty law but became recognised in customary international law. By 1945, a right to a territorial sea was recognised in international law and everything outside it as the high seas. The offshore oil industry has its origins in the United States and Mexico dating from early in the twentieth century. Wooden piers extended out into the Gulf to extract oil from under the so-called submerged lands within a 3-mile territorial sea. After 1945, operations extended well beyond 3 miles, and US President Harry S. Truman made the Truman Proclamations in that year. These unilaterally extended US resource rights over the seabed of the continental shelf while retaining a high seas designation for the waters above. Latin American states then made less modest declarations culminating in a 200-mile maritime zone in Chile in 1947.

The Hague Conference of 1930 had left the extent of the territorial sea unresolved. Work on this was started again by the International Law Commission in 1950, but already great changes in maritime practice had emerged. The historic emphasis on freedom of navigation and jurisdiction in the territorial sea had shifted to a focus on the sea as a reservoir of economic resources. The International Law Commission reported to the 1957 General Assembly of the United Nations. Drafts of texts were circulated to governments, revised and put to a diplomatic conference in Geneva in 1958. A two-thirds majority voting rule was adopted, and the 1958 Geneva Conference, the first United Nations Convention on the Law of the Sea (UNCLOS 1), reached agreement on the four Geneva Conventions on the Law of the Sea of 1958:

1   **Geneva Convention on the Territorial Sea and the Contiguous Zone** – which entered into force in September 1964. This confirmed sovereignty of the coastal state over a belt of sea adjacent to the coast, subject to the right of innocent passage.
2   **Geneva Convention on the High Seas** – which entered into force in September 1962. Coastal and landlocked states were to enjoy freedom of the high seas, including freedom of navigation, freedom of fishing, freedom to lay submarine cables and pipelines and freedom of overflight.
3   **Geneva Convention on Fishing and Conservation of the Living Resources of the High Seas** – which entered into force in March 1966. All states have the right for their nationals to engage in fishing on the high seas. The coastal state is recognized as having a special interest in maintaining the fish stocks

in waters adjacent to its territorial sea. The Convention requires all states to adopt all necessary conservation measures.

4   **Geneva Convention on the Continental Shelf** – which entered into force in June 1964. Coastal states exercise sovereign rights over the continental shelf for the purpose of exploring and exploiting its natural resources. 'Natural' includes mineral and other non-living resources of the seabed and subsoil plus living organisms belonging to sedentary species. The 'continental shelf' is defined as the seabed up to a point of a 200-m depth.

The 1958 Conventions still did not specify a width for the territorial sea, and a second Law of the Sea Conference (UNCLOS 2) was convened in Geneva in 1960. This dealt almost exclusively with limits to the territorial sea and zones in which coastal states could exercise preferential fishing rights, but the conference failed to reach agreement. A compromise proposal from the Canadian delegation for a 6-mile territorial sea and an outer 6-mile zone providing exclusive jurisdiction over fish stocks to the coastal state, subject to the 'historic rights' of other states, failed by a single vote to reach the necessary two-thirds majority. This 'six plus six' formula was subsequently adopted by 17 European states in the European Fisheries Convention of 1964. It was superseded by the European Economic Community (EEC) declaration of 200-mile fishing limits on 1 January 1977. The four Geneva Conventions failed to resolve competing claims or restrain some extreme claims over marine space, such as those from some Latin American states. The rate of increasing economic interest in the oceans had not been foreseen. Even fisheries, where national claims were already evident, had not been sufficiently provided for. In addition, decolonisation was gathering pace, and the needs and aspirations of the developing world were increasingly and firmly articulated. Work on a new convention began in 1967, and in 1970 the UN General Assembly adopted a Declaration of Principles:

> The seabed and ocean floor, and the subsoil thereof, beyond the limits of national jurisdiction . . . as well as the resources of the Area, are the common heritage of mankind . . . and shall not be subject to appropriation by any means by States or persons.

A new convention, the Law of the Sea Convention (LOSC) was adopted in 1982 at the third United Nations Conference on the Law of the Sea. It came into force in 1994 after receiving 159 ratifications, but several important and significant states were absent, notably the United States and the United Kingdom, although the UK later acceded in 1997. The new convention was mainly concerned with jurisdictional boundaries among nation states and giving clear definition to that which was lacking in the Geneva Conventions (Figure 4.1).

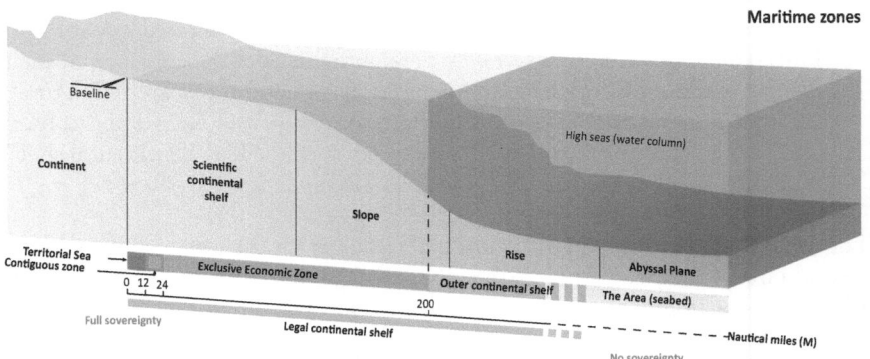

Figure 4.1 Jurisdictional zones following the adoption of the 1982 Law of the Sea Convention

Source: Riccardo Pravettoni, UNEP/GRID-Arendal, www.grida.no/resources/7923.

In addition to the adoption of a 12-mile territorial sea, two major new designations were introduced: the exclusive economic zone (EEZ) and The Area, which is the area of seabed beneath the high seas. The EEZ secures limited sovereign rights for coastal states but also imposes responsibilities. The main rights are to exploit economic resources, to construct islands and installations and to conduct research. The main responsibilities are to conserve and manage the resources and to control pollution. Navigation, overflight and the laying of cables and pipelines also remain open to the international community in common with the high seas. The Area, in accordance with a 1970 resolution of the United Nations, is declared to be the common heritage of mankind (UNCLOS Article 136), and no state or person may claim or exercise sovereignty or sovereign rights over any part it or its resources. Economic activities are to be for the benefit of mankind as a whole, and, in an unprecedented legal move, the Convention established the International Seabed Authority (ISA) to administer the governance provisions.

A number of last-minute changes were made to the Area to satisfy objections from those who had invested heavily in exploration for deep sea minerals. The most outspoken of the critics was the United States, who, with the UK and Germany (and hence the EU), did not become a signatory to the Convention in 1982. The UK and Germany have since acceded, but not the United States. The United States saw an opportunity to develop mining for deep sea minerals and did not wish to be subject to international arrangements for the licensing of The Area. Advisors to President Ronald Reagan argued that The Area was *res nullius*. After much wrangling, the dispute over this part (Part XI) of the Convention was postponed to a limited extent by a compromise agreement – the Implementation Agreement – although even this was not wholly accepted and remains under negotiation. However, seabed mining remains an industry in prospect only for the time being. Commodity prices have never been sustained at the point where the industry could be commercial. The mineral resources are there, and it has to be thought that one day it will work.

The 1982 UNCLOS has therefore set the jurisdictional zones for ocean governance that are firmly established in both treaty and custom. After over 2000 years of maritime law history, this is an achievement for the international community, although the scope for dispute, mainly about baselines and historic claims, is very high. Something like 130 cases are before the UNCLOS Article 76 Commission, which can rule on such disputes. A claim by the Republic of China for jurisdiction over most of the South China Sea is based on historic rights. At first sight, the claim appears to be in direct conflict with the EEZ rights of the Philippines, Vietnam and Malaysia, and it was ruled invalid in 2016 by the UN Tribunal. However, China rejects the ruling and continues to reclaim coral atolls in the area for military purposes. Such are the limits of international law that the only remedies, after negotiation and arbitration have failed, may be the opprobrium of fellow states followed by sanctions and, if serious enough, force of arms.

The UNCLOS is important because it is global and it sets the framework of law across the oceans and seas. Numerous other treaties, some concerning a few states and some equally global in intent, have been set in place mainly to deal with questions of pollution control and conservation (Box 4.1). An example of how these treaties work in practice is OSPAR (Convention for the Protection of the Marine Environment of the North-East Atlantic). OSPAR is the mechanism whereby 15 governments and the EU cooperate to protect the marine environment of the North-East Atlantic. It started in 1972 with the Oslo Convention against dumping and was broadened by the Paris Convention of 1974 to cover land-based sources and the offshore industry. These two conventions were unified, updated and extended by the 1992 OSPAR Convention. The new annex on biodiversity and ecosystems was adopted in 1998 to cover non-polluting human activities that can adversely affect the sea. The 15 governments (the contracting parties, or CPs) are Belgium, Denmark, Finland, France, Germany, Iceland, Ireland, Luxembourg, the Netherlands, Norway, Portugal, Spain, Sweden, Switzerland and the United Kingdom. The OSPAR Commission Secretariat facilitates the work of the CPs by supporting and organising an annual cycle of meetings together with intersessional work programmes, but the responsibility to implement and enforce the provisions falls to each member government. The CPs each have to report regularly to the Commission on (a) the legal, regulatory and other measures taken to implement the provisions of the Convention, including the prevention of or punishment for misconduct; (b) the effectiveness of the measures; and (c) any problems encountered. (See Box 4.1.)

Box 4.1   Key international treaties in ocean governance

BALLAST WATER 2004	The Control and Management of Ships Ballast Water and Sediments
BERN 1979	Conservation of European Wildlife and Natural Habitats
BONN 1979	Conservation of Migratory Species of Wild Animals
CITES 1973	International Trade in Exotic Species
ESPOO 1991	Environmental Impact Assessment in a Transboundary Context
ICES 1902/1964	Science and advice to support the sustainable use of the oceans
KYOTO 1992	Reduce greenhouse gas emissions
LONDON 1972	Prevention of Marine Pollution by Dumping of Wastes
MARPOL 1973/1978	Prevention of Pollution from Ships
RAMSAR 1971	Conservation and sustainable use of wetlands
Reg. Seas. BARCELONA	1976/1995 Protection of the Mediterranean Sea Against Pollution
Reg. Seas. BUCHAREST	1992 Protection of the Black Sea Against Pollution
Reg. Seas. HELCOM	1992 Protection of the Marine Environment of the Baltic Sea Area
Reg. Seas. OSPAR	1992 Protection of the Marine Environment of the NE Atlantic
Reg. Seas. UNEP-MAP	Implementation of the Barcelona Convention
SALVAGE 1989	Salvage of ships and goods at sea
UNCBD 1992	Maintain biological diversity
UNCLOS 1982	Law of the sea and nation state jurisdiction
UNESCO	United Nations Educational, Scientific and Cultural Organisation
UNFCCC 1992	Stabilise greenhouse gas concentrations in the atmosphere
Paris (COP21–2015)	Limiting global warming to less than two degrees Celsius

Similar arrangements to OSPAR are in place to manage each of these treaties. Governance at this level remains an interaction among sovereign states that voluntarily agree to be bound by the terms of the treaty. Below this level, we move to the domestic law of individual national regimes, which can directly bind the behaviour of individuals, companies and institutions. At this level, there is no voluntary agreement to be bound. The law is the law, and it has to be obeyed, but this opens up many new questions. How far should the law go? What regulations are acceptable to the wide range of interests and stakeholders? Who is going to decide, and just how will they do so?

For member states of the European Union, there is another level of maritime law. A sort of halfway house between international law and domestic law. Again, it is a body of law among sovereign states that has to be transposed into domestic law and regulations to take effect on individuals and businesses. However, it is more directly enforceable with a direct system of justice and fines (infraction proceedings) for countries not in compliance with the provisions of European law. Most of these laws affecting the marine environment are enacted in the form of directives. Many of the directives stipulate outputs that are to be achieved within set timescales but leave individual countries to decide on the detailed measures and regulations needed to achieve compliance. A significant exception is the Common Fisheries Policy (CFP), which is a regulation and directly sets fisheries management measures for all European states. The CFP is founded on the principle of a single EU Fisheries EEZ combining all the EEZs of the member states. A list of the principal EU maritime laws/policies and their administrative directorates are shown in Box 4.2.

Box 4.2 EU Maritime laws and policies

Law/policy	Directorate-General (DG)
Birds Directive – Special Protection Areas (SPAs) [BD]	DG Environment
Common Fisheries Policy [CFP]	DG MARE (Maritime Affairs and Fisheries)
Environmental Impact Assessment Directive [EIAD]	DG Environment
Integrated Maritime Policy [IMP]	DG MARE
Habitats Directive – Special Areas of Conservation (SACs) [HD]	DG Environment
Marine Strategy Framework Directive [MSFD]	DG Environment
Marine Spatial Planning Directive [MSPD]	DG MARE
Renewable Energy Directive [RED]	DG Energy
Strategic Environmental Assessment Directive [SEAD]	DG Environment
Water Framework Directive [WFD]	DG Environment

5 Governing ocean energy – governance at the working level

5.1 Background to national jurisdiction

Within national jurisdictions, maritime governance regimes and indeed any governance regimes in general are highly variable. There are many common themes, but every state has its own way of doing things in balancing the power of the government, the market and civil society. The power of the government may be distributed to several ministries and institutions or concentrated into single areas of responsibility. Powers may be held centrally or devolved to the local level. Every state has international treaty obligations to enact but has a choice of a myriad of measures and priorities available to achieve what it has agreed to. The European Union has a policy of subsidiarity in decision making, that is devolving decision making to the lowest level capable of making the decision.

Case studies are invaluable in these situations. They may be the only way of identifying emerging best practice in a rapidly changing framework of maritime governance at the national level. At this level, we are concerned only with the territorial sea and exclusive economic zones of the states concerned. It is in these zones that OE in the form of wave and tidal resources is set to be exploited and will be the focus of this section of the chapter. High seas areas may later be candidates for wave energy but are discounted for the time being. Governance at this level is also indivisible from marine planning, marine management, ownership, rights and community benefits. They feed off of one another. Every management or planning change depends on a governance regime or creates a new element of the governance framework, such as a new institution or a new enforcement/ monitoring agency.

The characteristics of wave and tidal energy devices make for a good case study of modern marine governance frameworks and a reasonable proxy for the challenges faced by other so-called Blue Growth industries. OE in the form of wave and tide is still an industry at an early prototype stage of development, but plans to accommodate it in coastal waters are well advanced in some areas, and full-scale trials give significant evidence of the impacts and interactions to be managed or governed. Typically these include:

- Large numbers of floating and seabed mounted devices in areas close to shore in the range of 0.1 km (kilometre) to 6.0 km, with significant visual impact and interaction, particularly with inshore fisheries and tourism/recreation;
- Proximity to shore a limiting factor on available landfall sites and increased risk of development in vulnerable or protected locations;
- Requirement for major linked terrestrial infrastructure at the coast, including ports, harbour services, generating stations and substations;
- Proximity to coastal areas of conservation, sites of natural beauty and marine protected areas.

Scotland arguably provides the most advanced case study of governance prepa-ration for OE and one of the most advanced marine governance frameworks in the world. Some might argue with such an assertion, but the priority given by the Scottish Government to wave and tidal energy as the driver for maritime governance and the establishment of the internationally recognised OE test site in Orkney (European Marine Energy Centre, or EMEC) add weight to the claim. Even before OE, the importance of offshore oil and gas introduced mod-ernised governance frameworks to Scottish waters dating back to the 1970s.

5.2 A Scottish case study

Scotland is a country within the United Kingdom and, for the present at least, the European Union. Scotland has a parliament in Edinburgh with substantially devolved domestic powers, while foreign relations and many financial instru-ments remain as reserved powers with the central UK Government in London. There is a third category of 'executively devolved' powers whereby the Scottish Government administers but cannot change UK-wide legislation. The Scottish marine area to the limit of the EEZ is large at nearly 600,000 km², which is about two-thirds the area of the whole UK waters and the fourth largest in Europe. Scottish waters are a major European centre for catch fisheries, aquaculture, off-shore oil and gas, conservation, tourism and recreation, and now for renewable energy mainly in the forms of wind, wave and tide. The Scottish TS and EEZ actually fall into two different jurisdictions. The TS is wholly under the control of the Scottish Government and the Marine (Scotland) Act of 2010. The EEZ outside the TS is the responsibility of the UK Government and the Marine and Coastal Access Act of 2009, but the administration of the area is executively devolved to the Scottish Government.

5.2.1 The marine acts

In common with many countries around the world, UK maritime governance has traditionally been a fairly haphazard and poorly organised area of admin-istration. Several ministries such as fisheries, transport, food, environment, energy and the like might be responsible for the sectoral aspects of marine administration, making integrated or concerted action slow and difficult. Infor-mal arrangements suited to particular communities and their interests under a kind of *mare nostrum* concept are common. Increasing development pressures, international obligations to control pollution, European law requirements, cli-mate change worries and Blue Growth ambition for future sustainable develop-ment demanded a change of approach. Parliamentary work on integrated laws and administration for the TS and EEZ started soon after the millennium and was completed in 2009/2010 with the two marine acts: the Marine and Coastal Access Act 2009, covering all of England, Wales and the Scottish EEZ outside of the TS, and the Marine (Scotland) Act 2010, covering the Scottish TS.

Take the Marine (Scotland) Act (MSA) as the example. With the stated aim to facilitate sustainable development and the mitigation of the effects of climate change, it introduced:

1   The establishment of Marine Scotland as the key government department dealing with all aspects of marine affairs in the Scottish TS;
2   Statutory marine spatial planning (MSP), comprising both a National Marine Plan by Marine Scotland and regional MSPs by appointed regional marine planning partnerships (MPPs);
3   Legal authority for the designation of marine protected areas (MPAs) and marine management areas (MMAs) in three classes: conservation, historic, and demonstration and research;
4   Streamlined development consent procedures working to a single point of contact (Marine Scotland) as the licensing authority (a 'one-stop shop').

In governance terms, the Act signals a determination of the government to bring clarity to the regime with a focused departmental responsibility and streamlined procedures to take sustainably based developments from application to approval in a speedy and systematic manner. It supports government strategy for marine renewables, including wind, wave and tide. The process tends strongly to central government control with consultation and advisory groups as the principal means of participation with stakeholders and communities. A significant exception is planning for aquaculture up to 3 miles from the coast. This remains the responsibility of the local authority under older legislation. The MPPs are delegated to draft the regional plans. Members of each MPP are anticipated to be the local authority together with a representative of Marine Scotland and other stakeholders. However, the MPP role is advisory only, and the actual planning power is retained by ministers. This is in contrast to terrestrial planning where authority (power) is substantially devolved to local councils with the possibility of an appeal to ministers in certain circumstances.

A further exception is the actual devolution of marine governance powers granted to the Shetland and Orkney Island Councils (SIC and OIC) in 1974 at the time of offshore oil and gas development in the North Sea. The Zetland County Council Act of 1974 gave SIC the power to control development in the TS around Shetland and special compulsory purchase powers for the oil terminal site at Sullom Voe. The powers enabled the council to negotiate a share of the North Sea revenues of the oil companies. Similar powers were applied in Orkney but to a smaller geographic area – just the area of Kirkwall Harbour and Scapa Flow.

5.2.2   Other institutions and organizations

Several other statutory institutions and organizations have maritime governance responsibilities in the TS and EEZ and may have direct legal powers to undertake certain actions (Table 4.1).

Table 4.1 Scottish marine governance organizations[14]

Organization	Responsibility
The Crown Estate Commissioners (TCE)	Administrators of UK The Crown Lands, which include the seabed in the territorial sea and certain activities in the EEZ. Legal duty (The Crown Estate Act 1961) is to maximise rents from the seabed (usually lease) with due regard to good management. Promoters of marine renewable energy.
Scottish Natural Heritage (SNH)	Statutory consultee in planning and authority on the implementation of the Birds and Habitats Directives and the network of nature conservation MPAs (MSFD/Marine Scotland Act 2010).
Historic Scotland (HS)	Statutory consultee in planning and authority on protection of archaeology and on built heritage, including wrecks etc. Advisers on the implementation of Historic MPAs.
Scottish Environmental Protection Agency (SEPA)	Government agency with responsibility for the implementation of the EU Water Framework Directive, which applies up to 3 miles from the coast. Statutory consultee in planning.
Local government councils	Land planning authority and planning authority for aquaculture up to 3 miles from the coast. The harbour authority in some cases.
Harbour authorities	Harbour management. Statutory consultee in marine planning.
Marine Planning Partnerships (MPPs)	Advisory bodies comprising central and local government members and other invited stakeholders. Under the Marine Scotland Act 2010, the MPPs can be delegated to draw up regional MSPs as advice to ministers.
Inshore fisheries groups (IFGs)	Advisory group comprising the fishing sector and environmental organizations. Responsibility to advise the minister on inshore fisheries management.
Coastal forums and partnerships	Voluntary bodies designed to encourage debate on coastal issues at a national level.

In addition, there are many non-governmental organizations (NGOs) and the general public who are keen to participate and influence the governance regime through formal consultation events, direct action and the work of their elected representatives. Elizabeth Kirk investigates the ways in which changes to marine governance regimes happen in practice.[15] Some are the result of catastrophic environmental events such as the depletion of the ozone layer or major oil spills. These prompt rapid changes to both laws and governance. However, where no catastrophe exists, the process is slow and is prompted by competing advocacy coalitions: "[P]olicy making occurs in a policy subsystem inhabited by several multi-actor advocacy coalitions which compete to influence policy in line with their beliefs". Science may be overlooked where it does not accord with the views of the dominant coalition. Kirk quotes the examples of Brent Spar (marine dumping), whaling and fisheries discards where NGOs captured public imagination and support to win changes not entirely in line with the science. Hathaway pointed out that the public tend to participate only when they understand the impact of the issues on their own lives.[16] Tippet et al. conclude that such an understanding comes at the operational stage rather than at the point of policy-making.[17] Hence the increasing public opposition to onshore wind farms as they proliferated and the low level of public response to OE consultations at the policy stage.[18]

6   Brexit and a window on maritime governance

In June 2016, the United Kingdom electorate voted to leave the European Union. The intricacies of the exit (or Brexit) negotiations are variously described as simple[19] to the most complex ever undertaken by a British or any government.[20] Over 40 years of EU membership have dominated British legislation and institutions, moving the country strongly towards rights-based governance. In engaging with European legislation, the courts were asked to construct a view of how to interpret diverse legal sources and applied 'modern constitutionalism' to a country, the UK, with no written constitution. Letsas writes: "In a rights-based constitutional democracy, no one is sovereign and no majoritarian decision, whether by people or by Parliament, is above the rule of law".[21] The constitutional issue of Brexit is the removal of individual rights in a constitutional democracy. Rights-based governance in marine space precedes even that introduced by membership of the EU and was extended by it and by international law. There are clear and obvious limits to the extent to which all that can be or should be unpicked.

Early preparations for Brexit provide a fascinating insight into the spider's web of interconnecting obligations and responsibilities towards governance of marine space and the resources it contains.[22] International law, European law, multilateral treaties, bilateral treaties and simple longstanding agreements to cooperate in fisheries management make any new arrangement fraught with difficulty if it is to propose a substantial change. Simply putting up a 'wall' at the EEZ boundary, as proposed by the UK Independence Party (UKIP) and others, will attract

retaliatory action from the UK neighbours and challenge international obligations for the management and conservation of fish stocks that are mobile over
several jurisdictions. Economically marginal but socially and politically important, the management of fisheries has been a major driver in the development
of maritime governance. The laws and associated management measures have
tried, with varied success, to recognise the mobility of the resource and to allocate catching rights with fairness. A principle that if a certain fish stock has a
nursery area in one jurisdiction and in adulthood wanders across several jurisdictions, then catching rights and conservation obligations should be shared to some
extent amongst all. The EU EEZ and the CFP are built on the principle that, for
fisheries purposes, there is one EU EEZ comprising those of all the member states.

As the Brexit negotiations attempt to unpick rights and obligations, they may
shine a light on why things are as they are and what needs to be changed. It is
only a 'may' because the scope for obfuscation is large, and the spider's web, as
detailed in Boyes and Elliott's 'Horrendogram', is dauntingly complex.[23]

7 Last words

The seas and oceans of the world have made a long and gradual transition from
being owned by no one (*res nullius*), with no private or public rights, to a position
where every square centimetre is situated somewhere in a governance framework.
In theory, *res nullius* no longer exists, having been replaced by communal property (*res communis*) or public property (*res publica*) and in, as yet, rare incidences
private property (*res privata*) – a slow process of enclosure. This is by no means
a settled state of affairs, as is evidenced by continuing debate about rights to the
resources of the high seas and the associated seabed (The Area). The UNCLOS
asserts these to be the "*common heritage of mankind*" and so communal property
under the jurisdiction of the International Seabed Authority (ISA). Several
countries object to this as a restraint on a fair return for their investment risk
in recovery technology and operations. Most prominent of these is the United
States, which has argued for continuing *res nullius* over seabed minerals under the
high seas. As a result, it has not signed or acceded to the UNCLOS 1982. It does
observe the general boundaries of the convention's jurisdictional zones, which
may now be regarded as Customary International Law. The case of seabed mining
is pertinent because it was the driving issue of principle in defining governance
for The Area. It remains a source of controversy, but seabed mining, thought to
be an industry on the cusp of a boom in 1980, has failed yet to materialise. It is
still held to be a promising prospect, but, to date, volatile commodity prices and
cheaper terrestrial sources of minerals have delayed its development. The ISA is
established as an entity, but it has yet to be tested under real pressure.

Governance of the oceans and seas has been built according to need: initially,
the need to conduct trade, manage fisheries and maintain national security across
national boundaries and areas of sovereignty. It is a dynamic and international
process that carefully ascribes rights and responsibilities to individual citizens

and the global community. It is changing all the time. Demand for maritime resources prompts technological advance to harvest them and investment in new industries. In a world hungry for resources and economic growth to sustain a burgeoning population, the pressures could be severe. Fisheries, shipping and offshore hydrocarbons have played seminal roles in building maritime governance. The so-called Blue Growth industries such as seabed mining, ocean energy, aquaculture, biotechnology and recreation promise to sustain the need for continuing development of governance, planning and management. The paucity of sovereignty and property rights at sea, the mobility of resources and pollutants and the capacity of nation states to argue about boundaries and rights all combine to make international and transboundary agreements the essential foundation of governance at sea. However, international law is a blunt instrument that applies only to the level of the nation state. It might influence but it cannot control the behaviour and actions of individual citizens or businesses. This has to come from the domestic law of the states concerned. In general, the domestic laws will reflect their international obligations, but the ways in which they are applied will vary in almost every case – no two countries will be the same. There will be fundamental differences in authority and rights, including human rights. The basic question is, "Who decides and how do they do it?" In the case of wave and tidal energy, this chapter has looked at the experience of Scotland because it is the most advanced world example of planning for large-scale deployment. This is explored further in Chapter 5.

A last word about public rights. Taking the United Kingdom as the example, there are public rights to navigate (Freedom of the Seas) and fish (subject to fisheries management measures). The rights at sea have to work in four dimensions: seabed, water column, sea surface and the air column above. This is explored further in Chapter 6 of this book. Maritime governance is a dynamic process that is perpetually 'under construction' as it develops and changes to reflect the needs of the time. For the present, it is another challenge to the emergence of OE industries, another uncertainty to be surmounted.

Notes

1 Boyes, S., and Elliott, M. (2014) Marine Legislation – The Ultimate 'Horrendogram': International Law, European Directives & National Implementation. *Marine Pollution Bulletin*, 86, pp. 39–47.
2 Kerr, S., et al. (2015) Rights and Ownership in Sea Country: Implications of Marine Renewable Energy for Indigenous and Local Communities. *Marine Policy*, 52, pp. 108–115.
3 Noble, T. (2003) Co-operating in Fisheries Management: Trials and Tribulations in Scotland. *Marine Policy*, 27, pp. 433–439; Wilson, J., Yan, L., and Wilson, C. (2007) The Precursors of Governance in the Maine Lobster Fishery. *Proceedings of the National Academy of Sciences of the USA*, 104(39), pp. 15212–15217.
4 Jones, P. J. S. (2014) *Governing Marine Protected Areas*. Abingdon: Routledge.
5 Plato (360 BCE) *The Republic*.
6 Hobbes, T. (1651) *Leviathan or the Matter, Form and Power of a Commonwealth*.

7 Dryzek, J. S. (1987) *Rational Ecology: Environment and Political Economy*. New York: Blackwell.
8 Jones, P. J. S. (2014) n. 4.
9 Denman, D. (1985) *Markets Under the Sea* (Film). London: British Film Institute.
10 Hobbes (1651) n. 6.
11 Johnson, K., Kerr, S. and Side, J. (2012) Accommodating Wave and Tidal Energy – Control and Decision in Scotland. *Ocean & Coastal Management*, 65, pp. 26–33.
12 Hansard (1973) *Zetland County Council Bill, Second Reading*. London: UK Parliament.
13 Portman, M. (2016) *Environmental Planning for Oceans and Coasts*. Switzerland: Springer.
14 Johnson, K., Kerr, S. and Side, J. (2016) n. 11.
15 Kirk, E. (2011) Marine Governance, Adaptation and Legitimacy. *Yearbook of International Environmental Law*, 22(1), pp. 110–139.
16 Hathaway, J. S. (1994) *The Clean Air Act and the Intermodal Surface Transportation and Efficient Act*. American Bar Association/Public Participation 28. Referenced in: Spyke, N. P. (1999) Public Participation in Environmental Decision Making. *Boston College Environmental Affairs Review*, 26(2), pp. 263–313.
17 Tippet, J., et al. (2005) Social Learning in Public Participation in River Basin Management. *Environmental Science and Policy*, 8, pp. 287–299.
18 Kerr, S., et al. (2013) Establishing an Agenda for Social Sciences Research in Marine Renewable Energy. *Energy Policy*, 67, pp. 694–702.
19 Farage, N. (2016) *UKIP EU Referendum Literature*, 23 June.
20 Farron, T. (2016) *Liberal Democrat EU Referendum Literature*, 23 June.
21 Letsas, G. (2016) Brexit and the Constitution. *London Review of Books*, 16 March 2016.
22 UK Parliament (2016) *Brexit: Fisheries*. House of Lords, European Union Committee, 8th Report of Session 2016–2017. HL Paper 78.
23 Boyes, S., and Elliott, M. (2014) n. 1.

Chapter 5

Marine planning

An ocean energy perspective

Kate Johnson and Glen Wright

I Introduction

It was long thought that the marine environment was not amenable to planning. Property rights, the core of the terrestrial planning system, were virtually non-existent in the marine environment, jurisdictions were vague, and the whole maritime area appeared as a single homogeneous mass of unstable waters without boundaries. The dominant and traditional maritime industries, fishing and shipping, had managed to coexist in relative peace for hundreds of years without formal planning processes to guide them.

However, technological innovation and the potential economic value of marine resources have created ambition for new industries, such as ocean energy (OE), aquaculture and seabed mining. Demands on marine space and ecosystems are expected to increase significantly as existing industries intensify and new industries enter the water. Fixed marine installations and users will ultimately be required to coexist with free-roaming industries like fisheries,[1] while vulnerable environmental and social systems increasingly demand greater control and planning. At the same time, advances in marine survey and spatial monitoring make planning marine space increasingly feasible and cost-effective.

While the future of these potential industries and their environmental impact is highly uncertain and too little is known at present to foster ideal policy responses, there is nonetheless a clear need for a pragmatic and flexible approach to marine management and planning. A range of spatial measures and sectoral policies have long been in place, but marine spatial planning (MSP) has more recently emerged as the leading concept for integrated marine planning in a modernised system of governance. A widely accepted definition is set out by Ehler and Douvere in the UNESCO guide to MSP:

> Marine spatial planning is a public process of analysing and allocating the spatial and temporal distribution of human activities in marine areas to achieve ecological, economic and social objectives that are usually specified through a political process.[2]

There is now a keen interest in MSP, but the practice is fledgling and the challenges are significant. Data on the marine environment and its users is often still insufficient to properly inform policymaking, and there remain complex questions of process, jurisdiction, governance and property rights. The discussion of the role and place of OE in MSP processes to date has generally been limited to calls for MSP to better advance and integrate OE,[3] assertions that MSP is crucial for the development of the OE industry[4] and consideration of how MSP applies to OE projects on a practical level.[5] There is also a growing thread of critical discussion in the literature relating to MSP in general[6] and to the challenges in relation to OE in particular.[7]

This chapter therefore provides a deeper discussion of the potential benefits and challenges for MSP in the OE context. Section 2 sets out the concept of marine planning, from its early origins in sectoral planning to the modern development of MSP. Marine planning is also considered in light of its more developed terrestrial counterpart. Section 3 considers the potential benefits of MSP for the OE industry, while Section 4 provides some brief case studies of MSP efforts, three of which have been driven by OE developments. Section 5 discusses the challenges of integrating OE into MSP processes, and Section 6 offers some concluding thoughts.

2 Marine planning

2.1 The genesis of marine planning

Despite its recent emergence, MSP is built on centuries of sectoral planning and zoning. Collisions at sea and the frequency of wrecks in the nineteenth century led shipping companies to define safe routes and shipping lanes for the use of their own captains, and these were ultimately codified by the Safety of Life at Sea (SOLAS) Convention of 1914. Additional shipping rules followed and were subsequently modified and incorporated into a 1960 version of the SOLAS Convention, administered by the International Maritime Organization (IMO). In the fisheries context, the International Council for the Exploration of the Seas (ICES), which came into being in 1902 after growing concern about the well-being of fish stocks in the North Sea, researches and provides statistics on spatially defined sea areas, an integral part of the management and planning of international fisheries.

Early initial steps towards increased planning were greatly accelerated with the advent of the offshore oil and gas industry in the 1940s, which, along with expanding fisheries, spurred enhanced claims for marine jurisdiction. For the first time, spatially defined platforms were fixed in the sea on a large scale, extending into deeper and more remote waters as the technology developed.[8] International concern about marine pollution and dumping introduced further elements to this emerging planning framework. For example, the OSPAR Commission works to develop policy and management for the protection and conservation of the marine environment in the North-East Atlantic,[9] and similar anti-pollution and -dumping conventions apply in several regions.

Nature conservation and ecosystem protection also began to play a more visible role in the development of planning and spatial definition at sea. Early large-scale marine reserves[10] led to the development of more modest networks of marine protected areas (MPAs) and marine conservation zones (MCZs), often targeted at specific habitats, species or features.

The need to mitigate climate change and the growing momentum of the so-called Blue Economy have brought new activities and complexities to this mix. Much of the new industrial development of the marine environment is close to shore and requires large areas of marine space. These industries introduce change of use (displacement or interaction with other activities) and change of non-use (destruction or modification of habitats and ecosystem services). This has provided a strong stimulus to move towards a more integrated and spatial approach to marine planning.

2.2 The emergence of contemporary marine planning

The multi-objective integrative process of MSP is a relatively recent development, having started in Europe around the turn of the century. It aims to answer three sets of questions:[11]

1   Where are we today? What are the baseline conditions?
2   Where do we want to be? What are the alternative spatial scenarios of the future? What is the desired vision?
3   How do we get there? What spatial management measures move us towards the desired future?

MSP aims to mitigate conflicts, assign priorities and promote cooperation and coexistence in pursuit of sustainable development. The movement towards MSP gained particular importance in Europe,[12] having started in the Benelux countries, Germany and Denmark. It has "rapidly become the most commonly endorsed management regime for sustainable development in the marine environment".[13] There is, as yet, limited practical experience with MSP. Functioning examples are sparse, and approaches are diverse in relation to national priorities.

The planning of marine space has been made technically possible by the development of geographic information systems and all the modern paraphernalia of marine survey. Data about oceanography, sub-sea topography, geology and marine biology are being gathered at unprecedented rates. Yet our knowledge of the sea remains patchy and incomplete. Scientific understanding of new maritime industries and their potential interactions with one another and the environment is quite poor, and this is especially true of OE, where the technology remains at the early stages of development and there is no commercial experience to call upon.

Emerging marine plans have therefore to deal with a great deal of uncertainty and a wide variety of base conditions.

2.3 From land to sea

Marine planning is compared to its more developed terrestrial counterpart, and integration of the two systems is an oft stated aim. Urban and city planning dates back millennia, but comprehensive and integrated national terrestrial planning systems are a product of the latter half of the twentieth century. In the UK, for example, land planning has its foundations in the social contract established between government and people after the ravages of World War II. Large swathes of British industry were nationalised at this time but land was not;[14] instead, a planning system was introduced for "subordination to the public good of the personal wishes and interests of private landowners".[15] In other words, the law would restrict the private rights of individual landowners for public benefit. The first UK Town and Country Planning Act was introduced in 1948. The original elements of this system remain:

1   Regional/local development plans setting priorities for the economy and community well-being
2   Spatial zoning with preferred patterns of land use
3   Development control requiring permissions for change of land use

Marine planning shares much in common with these ideas, but the basic tenets that influence their introduction are fundamentally different (Table 5.1).[16]

While the terrestrial planning system was developed to restrict the rights of private owners in an environment where nearly everything is owned by someone (*res privata*), marine planning addresses a space where there are no private owners. Marine space is largely either *res nullius* (owned by no one) or *res communis* (owned by the community). The UN Law of the Sea Convention (1982) assigns national jurisdictions and rights over certain resources in the exclusive economic zone (EEZ),[17] but at the national level, public and community rights are retained in these areas, most notably the right to navigate and the right to fish.[18] This question of ownership and rights is one of the key challenges to planning at sea (see Chapters 5 and 7).

Rights long exercised under regimes of *nullius* or *communis* have to be curtailed or reallocated if spatially defined new maritime industries are to thrive or if designated conservation areas are to be closed or partially closed to economic activities. Investors in spatially defined industries require a long-term right to occupy that space, often to the exclusion of others. While traditional industries like fishing and shipping are transient, many of the new wave of ocean industries are fixed. The terrestrial commons have long been enclosed, enabling the terrestrial planning system to control the actions of named individual owners, whereas the marine commons remains open, and a process of enclosure has barely begun.

Table 5.1 Key characteristics of terrestrial and marine environments

Terrestrial Environment	Marine Environment
Building blocks: • Individual locations dominated by single land uses • Absolute land ownership supported by law • Little public land • Enclosure of common land a historic fact • Private property rights held by individuals • Highly man modified	*Building blocks:* • Multi-user environment • Important common rights (e.g. fish, navigation) • Seabed managed by state on behalf of the public • High level of wildness • Quasi property rights being created
Development control: • 1943 origins of modern planning • Roots in modernist scientific approach • Recent shift towards planning through debate • Development plans with zoning supported by planning permission and development control as the planning key tools • Planning control limits the private rights of individual landowners • Significant role for local authorities and local priorities to influence decisions • Emerging mechanism for levering community benefits from renewable energy developments	*Development control:* • MSP a recent phenomenon • MSP driven by competing interests of environmental protection and economic development • Extreme reluctance to zone areas for specific uses (in UK and US), creating case-by-case decision making • EIA key to decision-making process • Decision-making power will be centrally driven by national priorities. • Limited scope for public involvement • Less opportunity to leverage community benefits
Conservation: • Traditionally urban and rural planning treated separately • Traditional conservation focuses on designation of small number of remaining sites that have high level of naturalness • Well established network of conservation designations and sites • Conservation designations driven by science • Relatively well understood environment • EU legislation increasingly important	*Conservation:* • Environment and development tension • Difficulty identifying MPAs in environment with high degree of naturalness • Conservation interests often highly mobile or dispersed • Specific locations of conservation value often impacted by off-site or transient pressures (e.g. pollution) • High levels of uncertainty in, distribution of habitats, nature of physical processes, and the extent and impact of human activity. • EU legislation increasingly important

The MESMA research project[19] examined conditions for MSP across 13 European countries with the aim of identifying the generic factors that could help build a common planning framework. The resulting framework was tested through 13 case studies,[20] and one important common factor that emerged in all areas was the presence of some existing policy and/or institution that could provide a starting point for an MSP process. Whether this was local fisheries plans or discrete MPAs, the basis for plans that could be developed through a series of iterations existed everywhere. A series of measures to deal with uncertainty are described in Section 5, but at its core lies a so-called Adaptive Planning Cycle that builds upon a starting point for a plan, however small.

2.4 Early MSP efforts

The Australian Great Barrier Reef Marine Park (GBRMP) and the Belgian EEZ are emblematic examples of the development of MSP. The GBRMP covers 344,400 km² and was created in 1975 by an act of the Australian Parliament to support the maintenance of ecosystems and biodiversity and the management of tourism, fishing and pollution. Concerns about potential phosphate mining, offshore oil, fishing, port developments and climate change were key drivers. The Park is managed by the GBRMP Authority, which with a current annual budget of approximately AUS$50 million (US$38 million). The GBRMP employs a zoning system, with the eight zones ranging from general use to preservation. A re-zoning exercise completed in 2004 resulted in the designation of so-called no-take areas covering about a third of the total area. Other changes over time have resulted in significant stakeholder participation and a recent Strategic Environmental Assessment (SEA) exercise initiated by UNESCO.[21] A feature of the GBRMP plans is continuous monitoring, evaluation and adaptation to changing circumstances.

In contrast to the GBRMP, the Belgian MSP was an early example of an attempt to implement multi-objective marine planning in an exceptionally crowded marine space. The Belgian coastline extends to a little over 80 km, with a marine area of only 3,600 km² to the limit of the EEZ. Work started on a non-statutory master plan around the turn of the millennium, resulting, after several iterations, in a statutory MSP promulgated by Royal Decree in March 2014. The plan was developed largely under the provisions of the Marine Protection Act 1999 led by the Ministry of the Environment, but a diverse range of responsibilities among government ministries led to the appointment of a Minister for the North Sea to coordinate government action. Existing uses included offshore gas, aggregates mining, shipping, fisheries, pipelines and aquaculture, and the main driver for the plan was the need to incorporate offshore wind energy and increased aggregates mining into this crowded space, while also meeting EU requirements for conservation. The small size of the marine area and the density of use lent itself to a completely zoned solution, such that in the resulting MSP, each activity is allocated a place.

3 The potential roles for MSP in the development of OE

Marine space planning has the potential to play up to five roles in developing OE: finding space for OE, reducing fragmentation, promoting sustainable development, easing conflicts, and providing information.

3.1 Finding space for OE

MSP may ensure that all marine activities, including new and developing sectors, are fairly allocated space to develop. Assuming that the process is holistic and inclusive, allowing trade-offs to be made between different ocean users, MSP has the potential to facilitate the integration of OE into a crowded marine environment. In addition, MSP may help policymakers to consider the variety of uses appropriate to the particular sea area in question, the relative values of different activities, the potential conflicts, and the suitability and sustainability of using different areas for different activities. In doing so, MSP should help to promote a mix of marine uses that are compatible with one another and the environment.

3.2 Reducing fragmentation

Marine governance arrangements are often characterised as fragmented, and nowhere is this clearer than in relation to OE. In most jurisdictions, OE is currently regulated through an array of pre-existing national, regional and international laws governing marine spaces (e.g. see Chapters 8–10 on consenting and environmental impact assessment processes). In this context, MSP has been promoted as a tool for both rationalisation and unification. MSP can, in theory, bring together existing governance instruments and bodies under the auspices of one process. Evidence from early national MSP efforts shows such an emphasis on MSP as a process aimed at increasing dialogue and understanding.[22] In order to combat fragmentation, integration would have to take place on several different levels, such as between legal instruments, different branches of government and different sectoral interests.[23] MSP may function as an umbrella under which different instruments of governance can be organised, thereby contributing to the achievement of a 'one-stop shop' or a similarly streamlined mechanism for adjudication and consent.

In a similar fashion, MSP may be able to alleviate issues relating to EIA and consenting, in particular by taking a future-oriented and strategic approach to balancing precaution and risk, providing flexibility and lending a level of predictability and consistency to the overall governance framework.[24] The geographical proximity of OE devices and the attendant onshore infrastructure raise the possibility that MSP may also be an appropriate mechanism to link emerging marine governance systems with terrestrial planning.[25] This will be increasingly important as OE projects begin to drive the development of additional harbour and port infrastructure, onshore facilities and grid extensions.

3.3 Promoting sustainable development

A sustainability-oriented process can ensure that OE development is done in a manner sensitive to the environment, while also acknowledging the environmental benefits of increased renewable energy deployment.[26]

3.4 Easing conflicts

MSP can improve stakeholder involvement by providing an open and transparent mechanism through which the interests of different sectors can be heard and reconciled.

3.5 Providing information

MSP may reduce costs of information, regulation, planning and decision making. MSP can improve efficiency, in terms of cost, by developing common approaches to the acquisition and dissemination of information, improving information provision and reducing duplication of effort. MSP can also provide regulatory efficiency by improving information exchange and providing a more certain environment in which regulatory decisions are made. MSP may be also expected to reduce regulatory and compliance costs.

4 Case studies in marine planning for ocean energy

Given that marine planning is an emerging discipline, there are not yet many well developed marine plans in existence, and few that have explicitly been driven by or have accounted for the development of a commercial OE sector.[27] This is changing in parts of Europe,[28] where many key jurisdictions now have programmes in place for the implementation of MSP. The promotion of offshore wind energy has been a strong driving force behind the development of national MSP frameworks in Belgium, Germany, the Netherlands and the UK.[29] However, a recent survey conducted for the International Energy Agency OE Systems programme (IEA-OES) found that of the 11 countries that participate, only four have formal MSP in place, three have no MSP and four have coastal management plans that can include marine and coastal uses such as shipping, fisheries and conservation.[30]

OE has nonetheless been a major driver of the development of marine planning in Scotland (UK), Oregon (US) and Nova Scotia (Canada). Scotland in particular has undertaken an extensive and ambitious programme of MSP driven in large part by OE.[31] The Pentland Firth and Orkney Waters (PFOW) plan in Scotland covers a region considered to be the most advanced testing ground for OE in the world and one of very few areas where advanced preparations are underway to support the commercial deployment of large-scale arrays of OE devices. It therefore forms the main case study in this chapter.

4.1 Oregon and Washington State, United States

The neighbouring states of Oregon and Washington lie on the Pacific coast of the United States, where a strong wave regime from the Pacific offers good prospects for the development of wave energy projects. The territorial seas of individual states in the United States extend to 3 nm from shore, giving Oregon and Washington marine areas of 2,600 km^2 and 5,200 km^2 respectively. Oregon established an Ocean Task Force to develop a Resources Management Plan (RMP) in 1990, and in 2008 work started on a Territorial Sea Plan (TSP), which aims to find suitable sites for the deployment of OE devices, protect fisheries, reduce conflicts and preserve ecological values. Washington State introduced specific legislation for MSP in 2010, driven by the prospects of offshore renewable energy and specifically wave energy. It is further supported by the federal Coastal Zone Management Act (CZMA) of 1972, which aims to preserve, protect, develop and, where possible, restore or enhance the resources of the coastal zone.

Oregon has amended its TSP to guide the siting of ocean renewable energy facilities. The relevant agencies conducted a spatial analysis of ocean uses and ecological resources through a public process to identify and allocate areas within the territorial sea that are appropriate for renewable energy development. In contrast to the Scottish MSP process, Oregon's Plan zones for different uses, ultimately designating 74% of Oregon's territorial sea as incompatible with OE and roughly 2% as "Renewable Energy Facility Suitability Study Areas".[32] The industry has expressed concern at what it sees as a 'negative approach' to MSP, excluding OE deployment where existing uses exist and focusing on constraints rather than on opportunities. Indeed, excluding areas from consideration for MRE development undermines one of the core benefits of MSP, which is that it allows for strategic planning and explicit trade-offs between uses, whether new or pre-existing.[33]

4.2 Nova Scotia, Canada

Canada holds considerable potential for the development of OE, especially in the Bay of Fundy (tidal) and the west coast of Vancouver Island (wave and tidal). Nova Scotia and the Bay of Fundy in particular appear to be best placed, with one study showing that 2.5 GW could be extracted from the most attractive site, the Minas Passage.[34]

The Nova Scotia Marine Renewable Energy Strategy is driven by this opportunity and need to develop renewable energy.[35] The strategy forms an integral part of the Province's clean energy framework, setting out the policy, economic and legal conditions for renewable energy projects in anticipation of commercial development and the establishment of a new industry. The objectives include delivering cost-competitive renewable energy to meet the need for more diversified and stable energy sources and developing an industry to provide opportunities to apply local knowledge and skills to serve global export markets. It sets out the elements for a phased and progressive approach to achieving a long-term goal of producing 300 MW of power from tidal stream projects. It includes initiatives for research, development assistance and regulation.

Comprehensive MSP is not yet under preparation in Canada, but key elements such as the creation of an integrated regulatory and consent regime is underway. The open ocean area adjacent to the Bay of Fundy, but not the Bay itself, is subject to the Eastern Scotian Shelf Integrated Management Plan (ESSIMP), which was completed in 2008 under the authority of the Canada Oceans Act of 1997. It is driven by increased competition for ocean space from sectors including shipping, offshore oil, wind energy, pipelines fishing, aquaculture and the military.[36] The Government of Nova Scotia has also conducted an SEA process in relation to OE development (further discussed in Chapter 10).

4.3 Orkney and Shetland, Scotland

The Northern Isles of Scotland (Orkney and Shetland) feature strongly in the emergence of OE technologies and advanced marine planning efforts. Orkney in particular is home to the European Marine Energy Centre (EMEC), which has become established as a global centre for the testing of full-scale prototype OE devices. Orkney was selected as the site for EMEC in 2002 because of the proximity of strong wave and tidal stream regimes combined with accessibility to sheltered port facilities and support services.

In 2008, The Crown Estate (TCE), the UK authority that manages the seabed, issued an invitation to companies to tender for seabed leases for wave and tidal energy developments in the area of the PFOW. The results were announced in 2010, and the spatial boundaries for development were established under this market process. At the same time a number of government actions were taking place, including the following:

- The enactment of the Marine (Scotland) Act 2010, which introduced statutory marine spatial planning, streamlined consenting procedures and powers to declare marine protected areas (MPAs).[37] The Act also created Marine Scotland as the government department responsible for marine affairs.
- The publication of a renewable energy road map establishing a target for all Scottish electricity to be generated from renewable sources by 2020, including 1.6G W of wave and tidal capacity installed in the PFOW.
- The creation of a Pilot Marine Spatial Plan for the PFOW region designed to test the MSP procedures under the new Act prior to roll-out to all the marine regions of Scotland. The PFOW was also designated as one of two Marine Energy Parks in the UK.

These were significant and ambitious changes, driven by optimism for a new industry and underpinned by the provision of considerable funding. The Scottish marine planning framework comprises a National Marine Plan (published in 2014), supporting an eventual suite of 11 regional marine plans, for which the PFOW plan is the pilot.[38] Regional plan drafting is to be delegated to local marine planning partnerships (MPPs).

The PFOW lies to the far north of Scotland, at the boundary between the Atlantic Ocean and the North Sea and is the primary shipping route between them. The Plan area measures about 12,000 km^2 and extends to the limit of the UK territorial sea around the Orkney Islands. The Plan boundary is at the limit of the territorial sea, which is also the jurisdictional boundary between Scotland and the United Kingdom as a whole. In general, the territorial sea around Scotland falls under the jurisdiction of the Scottish Government, although there are sectoral exceptions (e.g. offshore oil). The offshore area beyond the 12-nm territorial sea limit but within the exclusive economic zone falls under the jurisdiction of the UK Government but is administered by Scotland under a cooperation agreement.[39] The PFOW environment is relatively pristine and the subject of several environmental designations. The main traditional activities are community-based fisheries, shipping and ecotourism. In common with small islands everywhere, the sense of community 'ownership' of the seas around them is very strong. In practice, they have few maritime rights or control.

The Scottish Government had two main aims for the PFOW MSP:[40]

1   To facilitate sustainable development with strategic vision, policies and information; and
2   To develop a framework for integrating marine planning with terrestrial planning.

Work on the PFOW marine plan was started in 2009 and was divided into three stages: review of known data, a research programme to identify the most critical missing data, and preparation of the plan itself. The completed plan was finally published for consultation in March 2016, some four years later than initially intended.[41] The active OE operations and proposals in the area set the context for the plan. The awarding of seabed leases for OE in advance of MSP essentially defined the spatial boundaries, though the sites were still subject to the consenting process. It was decided early in the PFOW plan process that zoning was not possible. A similar conclusion was reached in the case of the Shetland plan and can be expected to apply across Scotland. The main reason for this is uncertainty about what the OE devices will actually look like, how they will work and their impacts.

The Marine Scotland planning team concentrated on a policy-based plan and guidance to prospective operators about possibly suitable areas for OE but without more detailed evaluation. The plan is policy based in that it sets general and sectoral policies about the criteria against which applications for development consent will be judged. It retains flexibility and options to face a number of scenarios. It supports the government's multi-use policy, even in protected areas. In other words, anybody can apply to do any activity in any part of the PFOW, but they know in advance what hurdles and requirements they are likely to face in gaining consent.[42] The completed Shetland MSP adopts a similar approach but goes a step further in evaluating possible development areas into a graded 'heat-map' (though there are no hard boundaries, such that this might be called soft zoning).

The three Scottish plans completed so far (PFOW, Clyde and Shetland) are all non-statutory pilots having the status of policy guidance to the statutory ter-restrial local plan. These marine plans therefore have some level of enforceability and are considered material considerations in any development proposal. Clyde and Shetland are expected to be made statutory, while the PFOW will be divided into two regions, Orkney and North Scottish Coast, before final publication as statutory plans.

5 The challenges of planning for OE

There are social, cultural and economic impacts to be assessed in the introduction of these new industries. The demand for private or quasi-private rights in previ-ously open and often relatively pristine areas of the marine commons is politically challenging. Planning and management are therefore an essential component of development.[43] The central challenge for the future of the OE industry is to move from this early developmental stage to a mature activity in a measured and sustainable way.

5.1 Uncertain outlook for the industry

Planning for OE is currently constrained by the uncertainty surrounding the ulti-mate shape of the industry. The timescale for commercial launch has slipped significantly, but much has been learned from the research and testing of the last ten years. OE technologies have yet to emerge as a commercial-scale industry, but so large is the prize and so consistent the ambition that an industry is likely to emerge. MSP processes will need to have a well balanced mix of certainty and flexibility in order to be able to support OE development and to adapt as the industry evolves.

5.2 Large demands for space

All maritime renewable energy installations require large areas of space. As the offshore wind industry has matured, the high output of individual turbines (now up to 10 MW) has led to increased spacing between adjacent towers in relatively remote locations, often outside 12 nm from the coast. This wide spacing has bolstered optimism that coexistence with fisheries, tourism and other activities is possible and may at some stage be agreed to. However, OE installations, as currently envisaged, differ substantially in character and location. The relatively small output of the first devices suggests that early commercial-scale projects will be fairly dense arrays in areas close to shore. Floating devices are further complicated by a network of mooring lines and anchors. Visual impact for wave and some tidal devices is therefore high, and coexistence with other existing activities is likely to be considerably more difficult.

This unprecedented demand for the exclusive use of large areas of marine space necessitates a more strategic approach to governance. OE devices enter an already crowded seascape, yet companies and investors need clear and long-term permission to occupy the space. Such enclosure of the marine commons is controversial and far from resolved (see Chapter 8).[44] Agreement and participation among stakeholders is one way forward (Chapter 13), but everyone is potentially a stakeholder in the marine environment, making the issue of occupation of marine space an issue of political significance.

5.3 Uncertain impacts and interactions

While considerable scientific research has been conducted into the potential environmental interactions of OE devices, there is still much uncertainty (see Chapter 9). This makes integrating environmental assessments and MSP processes difficult and requires MSP processes to remain flexible and adaptive as scientific knowledge matures.

5.4 Uncertain MSP processes

The nascent nature of MSP itself also presents challenges; it can be seen as all things to all people. For some, MSP is a broad planning instrument with little direct legal significance, while for others it is seen as a network of legally binding zones where only specified activities are permissible. The aims MSP should serve are also still controversial, with some agitating for the primacy of environmental protection and others arguing for MSP as a reconciliatory tool. Early experience with MSP suggest that so-called soft sustainability currently prevails, despite references to ecosystem-based management and the ecosystem approach.[45] For example, it has been argued that the EU has adopted a weak view towards sustainability and that MSP is, in fact, eroding existing environmental protections.[46] The development of MSP in practice has been far from homogeneous. EU efforts have been motivated in large part by economic goals, such as renewable energy targets, while much of the early support for MSP in the United States was generated by the academic and environmental advocacy communities, who saw MSP primarily in terms of marine conservation.[47]

5.5 Prioritisation of uses and coexistence

In Oregon, the MSP process excluded OE deployment in areas with existing users and focused on constraints rather than opportunities.[48] In contrast, Scotland's approach has been inclusive, developing separate policies for each existing activity in order to make considered trade-offs between users. However, to enable these trade-offs and to establish effective MSP, accurate and comprehensive data on the existing uses of the marine areas, their interactions and the condition of the environment is required. Prior to establishing priorities between uses, there has to be a clear view of how and to what extent different marine interests collide and how they might be alleviated by temporal and spatial allocation.[49]

The potential for coexistence of OE and other marine uses has been much discussed, particularly in relation to fisheries[50] and the potential for *de facto* marine protected areas.[51] However, successful coexistence is likely to be site specific,[52] and it is far from clear that OE devices, densely sited in nearshore areas, will ever be amenable to coexistence. As OE by its nature requires exclusive occupation of a specific marine space with particular resources, it may instead be preferable to 'zone' such uses, either within MSP processes or outside of them.[53] In most jurisdictions, consents are limited to a single use in a single location, so the possibility of multi-use licenses for larger areas would need to be considered.[54]

5.6 Resource allocation

There will be some difficulties in allocating resource access under MSP because developments may affect the availability of resources downstream. Unfortunately, the physics of wave/tidal resources and their interactions with devices are not well enough understood at present to factor this into planning. Flexibility will be needed to integrate additional knowledge as our understanding advances.

5.7 Sustainable development

It is important that the sustainability dimension of MSP is not lost in the rush to develop new resources. Sustainability criteria for MSP could be developed, possibly using similar criteria from other environmental governance instruments as a model.[55] This could be a step towards recognising the environmental benefits of renewable energies within legal processes and levelling the playing field with established marine activities.

6 Conclusion

The notion that the sea cannot be planned is looking increasingly outdated in the face of rapid technological developments. The push for a Blue Economy and plans for increasing industrial activities in the seas are bringing new spatially defined industries into close proximity with traditional free flowing activities such as fisheries and shipping, while enhanced knowledge of marine ecosystems and the services they provide implore policymakers to ensure sustainable development of these bountiful resources.

Several issues related to the management of marine space can be identified for further investigation:

1   A foundational question is whether the development of MSP is the best available option for integrating OE development into an increasingly crowded marine environment. Despite the rapid advancement of MSP, other tools, such as zoning, may be able to achieve the same aims in relation to OE and perhaps other unique cases. In Europe, pursuant to the MSP directive, states are obliged to pass MSP legislation and draft plans, though there are no strict substantive requirements. In other jurisdictions, there may remain substantial latitude to implement specific measures for OE where appropriate.

2   Assuming that MSP continues to develop as the preferred response, there are questions as to how best to integrate OE and other industrial uses into MSP processes. In any MSP process, the key concern will be how different activities will be prioritised. In the OE context, this has been achieved in different cases through an exclusionary approach and a plan-/policy-based approach. Further research is needed to develop good practice for MSP, particularly in relation to new and emerging industries. Regardless of the approach taken, issues regarding conflict resolution, coexistence and compensation will likely still arise.

3   Related to prioritisation is the question of coexistence. Clearly, further research is needed into a range of non-legal questions regarding feasibility of this, though subsequently there will be a need to develop appropriate legal and regulatory mechanisms to establish multi-use of sites.

Notes

1   Though note that Smith et al. (2012) argue that, although some industries are "mobile", in reality they are somewhat fixed through a range of existing sectoral spatial measures (e.g. shipping lanes for navigation). See Smith, H. D., Ballinger, R. C., and Stojanovic, T. A. (2012) The Spatial Development Basis of Marine Spatial Planning in the United Kingdom. *Journal of Environmental Policy & Planning*, 14(February 2015), pp. 29–47.
2   Ehler, C., and Douvere, F. (2009) *Marine Spatial Planning: A Step-by-Step Approach Toward Ecosystem-based Management.* UNESCO/IOC.
3   Thoroughfood, C. (2010) Marine Spatial Planning: A Call for Action. *Oceanography*, 23(March), pp. 9–10.
4   Ehler, C. (2011) Marine Spatial Planning – An Idea Whose Time Has Come. In: *Annual Report of the Implementing Agreement on Ocean Energy Systems.* Paris: International Energy Agency, pp. 96–100.
5   Wagner, A. (2010) *Report Defining the Criteria for Assessing National MSP Practices Affecting the Deployment of Marine Renewable Energy Sources.* Seanergy 2020; O'Hagan, A. M. (2012) *Marine Spatial Planning (MSP) in the European Union and Its Application to Marine Renewable Energy.* International Energy Agency Ocean Energy Systems Implementing Agreement website. Available at: www.ocean-energy-systems.org/ocean_energy/in_depth_articles/msp_in_the_european_union/

6 Frazão Santos, C., et al. (2014) How Sustainable Is Sustainable Marine Spatial Planning? Part I – Linking the Concepts. *Marine Policy*, 49, pp. 59–65; Flannery, W., et al. (2016) Exploring the Winners and Losers of Marine Environmental Governance/ Marine Spatial Planning: Cui bono ?/'More Than Fishy Business': Epistemology, Integration and Conflict in Marine Spatial Planning/Marine Spatial Planning: Power and Scaping/Surely, Not All Planning Is Evil?/Marine Spatial Planning – 'ad utilitatem onmnium'/Marine Spatial Planning: 'It Is Better to Be on the Train Than Being Hit by It'/Reflections from the Perspective of Recreational Anglers and Boats for Hire/ Maritime Spatial Planning and Marine Renewable Energy. *Planning Theory & Practice*, 17(1), pp. 121–151.

7 Wright, G. (2015) Marine Governance in an Industrialised Ocean: A Case Study of the Emerging Marine Renewable Energy Industry. *Marine Policy*, 52, pp. 77–84; Wright, G., et al. (2016) Establishing a Legal Research Agenda for Ocean Energy. *Marine Policy*, 63, pp. 126–134.

8 By the 1970s, the oil frontier was about 100 miles out into the North Sea at water depths of up to 300 m. Today, the frontier is far out in the Atlantic frontier at depths of up to 1,300 m. While the potential risks of such operations are considerable, the day-to-day impacts and footprints are relatively small and mainly of concern to the fishing industry.

9 Established by the OSPAR Convention of 1992 (with a history going back to 1967).

10 Such as Australia's Great Barrier Reef Marine Park (1976); the United States Florida Keys' National Marine Sanctuary (1980); and Ecuador's Galapagos Marine Reserve (1986).

11 Ehler, C. and Douvere, F. (2009) n. 2.

12 Douvere, F., and Ehler, C. (2009) New Perspectives on Sea Use Management: Initial Findings from European Experience with Marine Spatial Planning. *Journal of Environmental Management*, 90(1), pp. 77–88.

13 Flannery, W., et al. (2016) n. 6.

14 Kerr, S., et al. (2014) Establishing an Agenda for Social Studies Research in Marine Renewable Energy. *Energy Policy*, 67, pp. 694–702.

15 Report of the Expert Committee on Compensation and Betterment (1942) (commonly referred to as the "Uthwatt Report").

16 Kerr, S., Johnson, K. and Side, J. C. (2014) Planning at the Edge: Integrating Across the Land Sea Divide. *Marine Policy*, 47, pp. 118–125.

17 Thus, the rights of the State over its maritime zone might be considered as nationalized or *res publica* (owned by the public or state).

18 Todd, P. (2012) Marine Renewable Energy and Public Rights. *Marine Policy*, 36(3), pp. 667–672.

19 MESMA (n.d.) "Monitoring and Evaluation of Spatially Managed Areas". Available at: www.mesma.org

20 As varied as the huge area of the Barents Sea plan in Norway (1 million km²) to the tiny but densely used and well planned area of the Belgian Part of the North Sea (3,600 km²) and areas of the Mediterranean and Black Seas where planning has hardly started.

21 Related to its status as a World Heritage Site under the World Heritage Convention. See https://environment.gov.au/protection/assessments/strategic/great-barrier-reef

22 Stojanovic, T. A., and Farmer, C. J. Q. (2013) The Development of World Oceans & Coasts and Concepts of Sustainability. *Marine Policy*, 42, pp. 157–165.

23 Ehler, C., and Douevre, F. (2009) n. 2.

24 Day, J. (2008) The Need and Practice of Monitoring, Evaluating and Adapting Marine Planning and Management – Lessons from the Great Barrier Reef. *Marine Policy*, 32, pp. 823–831; Agardy, T., di Sciara, G. N., and Christie, P. (2011) Mind the

Gap: Addressing the Shortcomings of Marine Protected Areas Through Large Scale Marine Spatial Planning. *Marine Policy*, 35(2), pp. 226–232; Soininen, N. (2013) Planning the Marine Area Spatially – A Reconciliation of Competing Interests? *International Environmental Law-making and Diplomacy Review*, 2012, pp. 85–118.

25 Kerr, S., et al. (2011) The Integration of Land and Marine Spatial Planning. *Journal of Coastal Conservation*, 15(2), pp. 291–303. Additionally, terrestrial planning may also be used as a basis for understanding MSP. See Kidd, S., and Ellis, G. (2012) From the Land to Sea and Back Again? Using Terrestrial Planning to Understand the Process of Marine Spatial Planning. *Journal of Environmental Policy & Planning*, 14(1), pp. 49–66.

26 For discussion, see Foley, M., et al. (2010) Guiding Ecological Principles for Marine Spatial Planning. *Marine Policy*, 34(5), pp. 955–966; Frazão Santos, C., et al. (2014) n. 6.

27 Ehler, C. (2011) n. 4.

28 Douvere, F., et al. (2007) The Role of Marine Spatial Planning in Sea Use Management: The Belgian Case. *Marine Policy*, 31(2), pp. 182–191; Douvere, F., and Ehler, C. (2009) n. 12.

29 Qiu, W., and Jones, P. J. S. (2013) The Emerging Policy Landscape for Marine Spatial Planning in Europe. *Marine Policy*, 39, pp. 182–190.

30 O'Hagan, A. M. (2016) Maritime Spatial Planning and Marine Renewable Energy. *Planning Theory & Practice*, 17(1), pp. 148–151.

31 Marine Scotland (2010) *Pentland Firth and Orkney Waters Marine Spatial Plan Framework Regional Locational Guidance for Marine Energy: Final Report, Area*; O'Hagan, A. M. (2012) n. 5.

32 The Oregon Territorial Sea Plan is available at: www.oregon.gov/LCD/OCMP/pages/ocean_tsp.aspx. See Part Five: "Use of the Territorial Sea for the Development of Renewable Energy Facilities or Other Related Structures, Equipment or Facilities".

33 O'Neil, R. S., Geerlofs, S., and Hanna, L. (2012) Siting Wave Energy on the Oregon Coast, OES in-depth Articles Series. Available at: www.ocean-energy-systems.org/library/in-depth-articles/document/siting-wave-energy-on-the-oregon-coast-the-oregon-territorial-sea-and-siting-analysis-tools/

34 Karsten, R. H., et al. (2008) Assessment of Tidal Current Energy in the Minas Passage, Bay of Fundy. *Proceedings of the Institution of Mechanical Engineers: Part A: Journal of Power and Energy*, 5(222), pp. 493–507.

35 Nova Scotia (2012) *Marine Renewable Energy Strategy*.

36 Doelle, M. (2009) Role of Strategic Environmental Assessments (SEAs) in Energy Governance: A Case Study of Tidal Energy in Nova Scotia. *Journal of Energy & Natural Resources Law*, 27(2), pp. 112–144.

37 See Johnson, K., Kerr, S., and Side, J. (2012) Accommodating Wave and Tidal Energy – Control and Decision in Scotland. *Ocean & Coastal Management*, 65, pp. 26–33.

38 Other Scottish MSP pilots were under preparation in Shetland and Clyde, but the PFOW plan was unique in Scotland in testing the new statutory procedures and unique in the world as a plan facing up to the implications of real proposals for the immediate development of large-scale OE projects.

39 There are exceptions for some activities – e.g. offshore oil and gas, which fall to the jurisdiction of the central government wherever it is situated in UK waters.

40 Marine Scotland (2013) *Pilot Pentland Firth and Orkney Waters Marine Spatial Plan: Planning Issues and Options Consultation Paper*. Edinburgh: Marine Scotland.

41 Marine Scotland. (2016) *Pilot Pentland Firth and Orkney Waters Marine Spatial Plan*. Available at: www.gov.scot/Publications/2016/03/3696

42 ABP Marine Environmental Research Ltd. (2012) *Marine Scotland Licensing and Consents Manual, Covering Marine Renewables and Offshore Wind Energy Development*. Edinburgh: Marine Scotland.

43 Johnson, K., Kerr, S., and Side, J. (2012) n. 37; O'Hagan, A. M. (2016) n. 30.
44 Todd, P. (2012) n. 18; Kerr, S., et al. (2015) Rights and Ownership in Sea Country: Implications of Marine Renewable Energy for Indigenous and Local Communities. *Marine Policy*, 52, pp. 108–115.
45 Frazão Santos, C., et al. (2014) n. 6.
46 Qiu, W. and Jones, P. J. S. (2013) n. 29.
47 Gopnik, M., et al. (2012) Coming to the Table: Early Stakeholder Engagement in Marine Spatial Planning. *Marine Policy*, 36(5), pp. 1139–1149.
48 O'Neil, R. S., Geerlofs, S., and Hanna, L. (2012) n. 33.
49 Ehler, C. and Douevre, F. (2009) n. 2.
50 Rodwell, L. D., et al. (2012) Fisheries and Marine Renewable Energy Interactions: A Summary Report on a Scoping Workshop for the Marine Renewable Energy Knowledge Exchange Programme (MREKEP); de Groot, J., et al. (2014) Investigating the Co-existence of Fisheries and Offshore Renewable Energy in the UK: Identification of a Mitigation Agenda for Fishing Effort Displacement. *Ocean & Coastal Management*, 102, pp. 7–18.
51 See, e.g. Yates, K. L., Schoeman, D. S., and Klein, C. J. (2015) Ocean Zoning for Conservation, Fisheries and Marine Renewable Energy: Assessing Trade-offs and Co-location Opportunities. *Journal of Environmental Management*, 152, pp. 201–209; Kyriazi, Z., Maes, F., and Degraer, S. (2016) Coexistence Dilemmas in European Marine Spatial Planning Practices. The Case of Marine Renewables and Marine Protected Areas. *Energy Policy*, 97, pp. 391–399.
52 Christie, N., et al. (2014) Co-location of Activities and Designations: A Means of Solving or Creating Problems in Marine Spatial Planning? *Marine Policy*, 43, pp. 254–261.
53 There has already been some debate as to the relationship between zoning and MSP that may be relevant to the OE sector and other industrial users. See, e.g. Ehler, C., and Agardy, T. (2013) *Online Debate: Does Marine Spatial Planning Need to Involve Ocean Zoning to Be Effective?* Open Channels. Available at: http://openchannels.org/chat/online-debate-does-marine-spatial-planning-need-involve-ocean-zoning-be-effective-bud-ehler-and
54 Stuiver, M., et al. (2016) The Governance of Multi-Use Platforms at Sea for Energy Production and Aquaculture: Challenges for Policy Makers in European Seas. *Sustainability*, 8(4), p. 333.
55 See Bosselmann, K., and Engel, P. (2008) *Governance for Sustainability: Issues, Challenges, Successes*. Gland, Switzerland: IUCN, IUCN Environmental Law Centre.

Chapter 6

Mare reservarum

Enclosure of the commons and the evolution of marine rights in an era of ocean industrialisation

Sandy Kerr, Kate Johnson, John Colton, Glen Wright and Stephanie Weir

I Introduction

Generally speaking, private property rights end at the beach. Move across the tide line, and the ability to undertake any activity or use the marine space becomes a matter of jurisdiction and rights – a right of free navigation, a right to fish, a right in the exclusive economic zone (EEZ) to build installations or islands and exploit certain resources. The seas are an international arena in their entirety. Even the territorial sea, which grants almost complete sovereign rights to the adjacent state (innocent passage being the exception), is underpinned by international agreement and law, and with jurisdiction come responsibilities – a responsibility to manage, a responsibility to conserve and a responsibility to allow free access in the EEZ and innocent passage in the territorial sea. Ownership, at first sight, does not enter into the equation, but the emergence of actual or quasi-property rights at sea is challenging this assumption.

Maritime governance and maritime planning were discussed in Chapters 4 and 5 of this book, and the complex question of rights and ownership is alluded to in both. Chapter 4, on governance, explored the terms *res nullius* (owned by no one) and *res communis* (owned by everyone) and their different meanings for rights. If *nullius* applies, then nobody owns it, and anyone may claim it without an infringement of rights. But if *communis*, then everyone owns it, and everyone's rights are infringed by an individual claim. International law increasingly asserts *communis*, including for the seabed beyond national jurisdiction (called The Area in the UN Convention on the Law of the Sea (UNCLOS), which is covered by an implementing agreement to UNCLOS and managed by the International Seabed Authority (ISA).

For centuries, the main maritime activities of shipping and fishing have worked well within the rights regime, and the question of ownership hardly arose. They are both free-ranging transient activities, and the right to roam was the essential component of their industries. Certainly there were disputes and even the occasional war. More recently, two trends have begun to change this, slowly enclosing the marine commons or at least significant parts of it.

The first of these is management of the wild fisheries. Pressure on stocks is great, and it has been seen as necessary to decide who can fish where and for what. National and regional quotas for specific species have been set and enforced. In many regimes, these have been made transferable, and a market has developed in tradable fish quotas, a private property right in a marine resource. The implementation of this policy has ideological roots in the belief that privately owned stocks will be better managed than common stocks.

The second trend is the increasingly rapid development of maritime industries that are fixed in location. The owners of these businesses need a right to occupy sometimes in perpetuity and in all cases for a very long time. The industries include such activities as offshore oil platforms, aquaculture installations, offshore wind farms, seabed mining and ocean energy (OE). Advances in technology make their spread commercially viable, and many governments are promoting the development of the so-called Blue Economy, focused on creating jobs and economic growth while ensuring sustainability. Clearly such fixed infrastructure obstructs fisheries, shipping and the right to roam in varying degrees. In most cases, such industries obtain a right to occupy in the form of a seabed lease from their government. These are usually tradable, and again a market has developed in site licenses for aquaculture or energy installations. Quasi-private property rights at sea have therefore been created.

Promoters and regulators of OE and other spatially defined marine industries face a dilemma. They have a need to lay foundations, create structures and moor vessels and devices that inevitably obstruct navigation and fisheries. At the same time, they have a duty to preserve freedom of navigation and fisheries. The dilemma is complicated by the existence of different rights in four dimensions at sea: seabed, water column, sea surface and the air above. On land, only two dimensions have to be considered, land and air. A series of pragmatic measures have been enacted to help deal with real or prospective conflicts at sea, but a comprehensive solution is not yet available and is further complicated by uncertainty of what is required. What will these technologies look like, and what are their needs? It therefore becomes a case-by-case approach dealing with issues as they arise.

In any case, the expansion of private property rights or quasi-private property rights at sea is a departure from historical practice. The logical extension of these trends is that the marine commons will gradually become enclosed and that the old doctrine of Freedom of the Seas will be severely curtailed. New rights will have to be established in support of new activities and old rights modified or extinguished. This will not come easily.

This chapter explores these issues in more detail and offers some reflections, using the OE industry as a background for discussion. Section 1 looks at OE and public rights at sea, with particular reference to the UK, where commercial OE deployments are on the horizon. Section 2 examines fisheries and the consequences of the policy to privatise fish stocks through tradable quotas. Section 3 is a case study of the Shetland Islands and the measures enacted there to ease the arrival of North Sea oil into the marine rights regime around the islands. Section 4 investigates the development of the wider Blue Economy and the effects of a right of occupancy on existing regimes. Section 5 reviews indigenous rights with particular reference to Canada and Australasia. Finally, Section 6 provides a summary and considers some future directions with respect to OE development.

2 Towards quasi-property rights: fisheries and tradable quotas

Rights-based fisheries management is a relatively new concept in the history of ocean governance. As has been mentioned, the marine environment was treated as effectively open access for centuries, and this philosophy carried over to fish stocks; it appears there was very little regulation of any ocean fishery until the mid-nineteenth century.[1] Fisheries are arguably the most recognisable 'commons', eternally referenced as the epitome of overexploitation, with selfish users and rapidly deteriorating fish stocks.[2] Many have attributed failing fisheries to a lack of adequately defined property rights in both the stocks and their marine environment.[3] In much the same way as has been recognised in numerous other resources, introducing property rights and ownership-like powers to fishing has been lauded as an effective method for strengthening individual responsibility over the resource and consequently improving resource sustainability.

There are two primary methods of management that are common to most fisheries, both of which can utilise property rights to different degrees. Input (effort) controls incorporate rights through the use of access licenses or days-at-sea limitations, whilst output controls include the more recognisable quotas and catch limits. The structure of a regulatory plan will be specific to the fishery, incorporating a variety of input and output controls to allow for the vast array of unstable externalities within a reasonably unpredictable resource. For non-migratory species, using rights to limit output is a relatively simple task.

Territorial use rights in fisheries (TURFs) are used for species such as crustaceans, which exhibit low migratory behaviour, and thus operate on a fixed-space basis.[4] TURFs allocate fishers the exclusive right to harvest within an established marine region. However, highly mobile species are a far more difficult resource to manage correctly, and so a range of instruments have been developed.

A major innovation was the introduction of total allowable catches (TACs) in the 1970s, which refer to the allowable catch for an individual species, calculated by way of historic catch records. Within the EU, TACs are split into national quotas for each member state. These national quotas are then further spread through fishing regions and producers' organizations, down to individual vessels. Individual quotas, in particular individual transferable quotas (ITQs), are perhaps the most universal instruments that employ property rights in fisheries management. In theory, individual quotas allow for more sustainable fishing practices by committing fishers to pay more for the ability to catch more. This in turn should ideally ensure that fishers think more closely regarding the health of their stocks, as they have invested money in their long-term future. It also decreases intense rivalry between fishers that could otherwise cause overexploitation.

Quota systems bequeath fishers certain ownership-like powers: the power to use a resource, the power to extract from it, and the power to sell it. Possessing ITQs essentially provides the fishers an exclusive right to harvest their allocated tonnage and land it when they choose.[5] Furthermore, a key characteristic of ITQs is their transferability; unlike rigid licenses, ITQs can be bought and sold amongst fishers. The introduction of quotas has created multiple competitive markets that fishers act within; these markets should adequately represent the future health, both biological and economic, of a fishery. Additionally, the ability to buy, sell and occasionally lease price-based regulatory instruments theoretically increases the efficiency of the fleet, being more advantageous for less efficient vessels to sell quota than actually fish. To what extent these instruments constitute private property is perhaps outside the scope of this chapter.

Ownership of a quota should not be conflated with ownership of the fish themselves; an ITQ gives an individual a right to hunt, without assigning property rights to the physical resource.[6] It can be argued that most catch-share systems occupy a space somewhere between open access and privatised regimes. Nevertheless, it is undeniable that the introduction of such methods is encouraging the ongoing transition of fish stocks away from open access regimes. Whilst UNCLOS resulted in the enclosure of 90% of fisheries yield into state jurisdiction,[7] rights-based instruments like ITQs have meant that the enclosure of fish stocks has moved away from a purely spatial basis.

Rights-based instruments are not a panacea for the problems of managing highly mobile resources that frequently cross national boundaries. Detractors of quotas argue that multiple problems have arisen because of such schemes. ITQs, for example, have been accused of increasing incidents of high grading, the practice of discarding smaller or less valuable fish in order to maximise the returns from the fishers' quota.[8] Grant et al. suggest that ITQs and similar methods of enclosure have resulted in the "erosion of community" and the "proletarianisation of fishermen";[9] similarly, Cardwell and Gear state that the "high price of entry" into an ITQ-regulated fishery jeopardises efficient renewal from new fishers.[10] However, acting as sharply focused policy instruments, quotas and their kin have had largely favourable effects on limiting overexploitation, demonstrating that implementing stronger property rights onto a resource can prove beneficial to its sustainability.[11]

3 Shetland and the arrival of offshore oil

The dramatic discovery and exploitation in the 1970s of oil and gas resources in the UK sector of the North Sea created immediate challenges to the existing regime of marine ownership and public rights. The prospect of self-sufficiency in energy was a godsend for the UK, which at the time was reeling under the combined onslaught of Middle East oil boycotts, oil price rises and striking coal miners. The discoveries were in deep water and a long way, over 150 km, from the nearest land, which in many cases was Shetland.

The Shetland Islands are located around the sixtieth parallel and are the most northerly part of the UK. They lie at the dividing line between the Atlantic Ocean and the North Sea, over 1,000 km north of London, 500 km north of Edinburgh and 200 km north of the Scottish mainland. Before oil, the islands' economy was small but successful, based on fisheries, crofting (micro farming) and craft industries such as knitwear. Fishing was and is the most valuable of these traditional industries, and Lerwick, the capital, is now the second largest fishing port in the UK. The population in 1970 was a little over 17,000 people.

Development of the offshore resources near Shetland moved the technology into conditions never before attempted (i.e. deep water and rough weather), but the pressures for rapid development and delivery of the oil were great. It was quickly realised that the only practical solution for the northern oilfields was by pipeline to land and stabilise the crude oil in Shetland before shipping to markets around the world. The prospect of the oil-related industries clearly promised considerable opportunity but with threats at the same time. This might be something entirely out of the control of an island community that was accustomed to managing its own affairs. Soon after the announcement of the landing of oil in the islands, the speculators arrived buying up options to purchase land suited to a terminal site. The deep water fjord of Sullom Voe was the centre of attention. The local authority for the islands, Zetland County Council (ZCC), was starting a transition to become the Shetland Islands Council (SIC) under local authority reforms.

The first SIC Chief Executive Ian Clarke and the convenor of ZWW George Blance considered measures to benefit from the economic boost that the oil would bring but at the same time keep some control of events and protect the islands for the future. The life of the oilfields was thought at that time to be less than 25 years. The short but dramatic impacts could decimate the small but successful traditional economy and then disappear, leaving no long-term benefits. They decided to seek an Act of Parliament, a Private Bill, granting the council a series of powers allowing them to guide events and not just to be swept along by them. They gained the support of their Member of Parliament, Jo Grimond, who was a highly skilled orator and parliamentarian who was previously the leader of the Liberal Party. A bill was tabled in 1972 and by April 1973 and its second reading, it had more or less achieved the final content that was passed in November before coming into force in April 1974 as the Zetland County Council (ZCC) Act 1974. The passage of the bill had not been easy, and it had faced serious opposition.

The provisions of the Act are fivefold:

1   Special powers for the Council to compulsorily purchase land needed for the oil terminal. The Council were anxious to interrupt the speculative activity and to have power to guide the oil companies to a site that suited the community. They were also able to insist that the companies shared one site in

joint venture. As owners of the site, the Council were able to realise rents as well as local taxes.

2   Development control and environmental management powers in the marine environment (local authority jurisdiction normally ends at the beach). The Act grants the SIC developmental and environmental control powers to the limit of the UK territorial sea around the islands.

3   Appointment of SIC as the harbour authority for the oil terminal port and surrounding waters with power to own and operate tugs and towage.

4   Financial powers giving the Council the authority to borrow money to invest in port equipment and infrastructure.

5   Agreement that revenue from terminal rents and taxes, harbour operations, and other agreements with the oil companies remain in Shetland for the benefit of the community.

In moving the second reading of the Bill in parliament, Jo Grimond reasoned:[12]

> The Bill, which deals with the situation that has arisen owing to the discovery of oil off Shetland, may well point the way to better methods of controlling developments consequent upon oil discoveries in other parts of Britain. But the Bill has arisen because of the Shetland situation, and the Shetland situation is to some extent a special one. I believe it is essential that the impact of oil on a rather unique island community such as Shetland – an impact which may be extremely serious but which certainly will not go on forever and may be over after 50 years – should be controlled carefully. We must not sacrifice the long-term future of the community. Control should be exercised with several aims in view. The first is the well-being of all Shetlanders and all their descendants. The second is the maintenance of their traditions, which are largely Norse in origin and which differ from the traditions of much of the rest of Britain. The third is the preservation of the best land of Shetland – here I have in mind both its agricultural value and its beauty – and the protection of Shetland's existing fisheries, agriculture and industry. We should also arrange for the retention in the islands of a fair share of the income which will arise from these discoveries. We should ensure that this income can be used for the general well-being and for such purposes as making good if pollution should take place and the ultimate rehabilitation of sites which may be used and then discarded. The fifth aim is that service and other industries connected with the oil should be introduced in an orderly manner and at a speed and on a scale which the islands can absorb without disruption.

They are stirring words and just part of a long oration that carried the day, rather against the odds. The many objectors to the Bill argued first that the precedent was unacceptable. If passed for Shetland, then similar powers would have to apply to every local authority in the UK. In fact, 40 years on, the legislation is unique to Shetland except for the neighbouring Orkney Islands where a similar response to the oil industry is reflected in the Orkney County Council Act 1974. The same powers apply in Orkney but to a more limited geographic area around Kirkwall and Scapa Flow. A second objection believed that a small authority like Shetland could not possibly manage the magnitude of the task. Again, with hindsight, the objectors have been proved wrong, and the islands have benefited both economically and environmentally from the powers they were granted. The powers enabled the council to negotiate both disturbance payments and a share of oil revenues with the companies concerned. It is difficult to determine the value of the benefit up to the present day, but it is believed to exceed £500 million of capital invested for the benefit of community services and infrastructure in Shetland. Environmental management is shared with the council, the industry· and NGOs in SOTEAG (Shetland Oil Terminal Environmental Action Group). A final objection was made on the grounds of market intervention. The Act, it was said, represented unnecessary intervention in markets and was a misuse of state power. This objection counts more today than it did in the 1970s. After more than three decades of market-oriented economic policy, the more socialist ethos of the past is easily forgotten.

Clearly there were special circumstances that brought the ZCC Act into being, and the industry concerned differs markedly from the emerging industries of the new millennium. The oil industry was rich, and the commodity had an established market value well in excess of the cost of recovery. Companies were able to pay to ease challenges to a rapid and efficient operation from recovery to market. Renewable energy has no such excess of value to fund problem solving and is dependent on subsidies. And the economic times have changed. It is difficult to imagine such an interventionist arrangement passing the countenance of governments today. However, the Shetland model has worked and is working. It is one answer to the question of rights and ownership of the Blue Economy where a whole community holds to a tradition of the seas around them as being owned by all of them.[13]

4 Ocean energy and public rights in the UK

Developers of OE technologies need to have certainty of their rights to occupy the space from which they can conduct their business. Areas of seabed, water column, sea surface and even the air mass above might be required for their exclusive use, or perhaps their shared use in defined circumstances. However, unlike on land, there are no private property rights or freehold that they can purchase to ensure this certainty.

International law lays down rules in support of the freedom of navigation and fisheries. Nation states have diverse sets of rules, some preceding international law and others being shaped by it. These vary widely among countries, though the UK is discussed here as an example of a jurisdiction with an advanced marine economy, currently hosting several new developments requiring fixed marine space, including wave power, tidal stream, offshore wind, offshore oil and gas, and aquaculture.

The British monarch at one time owned all the land. Over centuries, parcels of land were awarded to loyal supporters (or to buy off enemies). Divided and divided again, these led to the pattern of freehold land ownership seen today. Any leftover parcels became known as the Crown Lands, and this included the foreshore and the seabed in the territorial sea. Although still referred to as Crown Lands, they are now, in effect, public lands, administered under agreement between the State and the Crown in return for taxpayer support for the costs of the monarchy.

The Crown Lands are administered on behalf of the public by the Crown Estate Commissioners under The Crown Estate Act 1961. The Commissioners are under a legal duty to obtain the best income possible from the Estate, and they have an interest in promoting rents from seabed activities. Revenues are passed to the Treasury. The areas of seabed outside of the territorial sea but within the EEZ are not Crown Lands but are treated similarly for the purposes of a right to occupy. Under the Continental Shelf Act of 1964, The Crown Estate (TCE) manages rights to natural resources on the UK continental shelf. Significantly, this excludes oil and gas, which are separately administered. More recently, the Energy Acts of 2004 and 2008 have added to TCE's portfolio the management of the rights to generate electricity from renewable energy resources on the continental shelf and the underground storage of natural gas and carbon dioxide.

In the UK, it was the courts that were first called upon to interpret ancient maritime rights and existing laws as they bumped up against the modern era. Even here, the cases are quite old and very few, but their interpretations and subsequent rulings set important and current precedents. Many concern challenges to decisions of TCE over permissions and rents for things like fixed moorings and jetties, others between groups of fishers.

English and Scottish authority over the seabed was most recently reviewed by the courts in the case of the *Crown Estate Commissioners v Fairlie Yacht Slip Ltd* in 1979. The case agreed with the right of the Crown to alienate (sell or lease) the seabed but with the important condition that the Crown cannot make a grant that might "in any way interfere with the uses of navigation or with any right in the public". On appeal, this was clarified to say that "the right of navigation is not to be regarded as a right to sail in every square inch of the surface of the sea". In other words, the public rights are infringed only if there is "material interference". These words leave ample scope for doubt and the need for definition, perhaps in future court cases. Nothing so far changes the statement in an earlier case (*Denaby and Cadeby Main Collieries v Anson* 1911) that:

> the ownership of the seabed is for the benefit of the public, and cannot be used in any manner so as to derogate from, or interfere with, the right of navigation which belongs by law to the subjects of the realm. The right of the pubic to navigate in navigable waters is a right to which the right of the Crown in the subsoil is subservient.

A renewable energy developer obtaining a right to occupy (through sale or lease) an area of seabed in the territorial sea or a renewable energy zone in the EEZ therefore takes on the obligation to preserve the public rights of navigation and fishing in the waters above, unless there is a change in the law. The UK Government, in line with international law, has made limited changes to give developers the exclusive rights they need and relief from public rights in fairly narrow circumstances. The Energy Act 2004 (Section 95) allows for safety zones to be created where renewable energy arrays of devices are to be consented and constructed, extended, operated or decommissioned. Vessels can be excluded from the area. In the case of wind turbines, this is specified at a radius of 500 m from each turbine during construction and 50 m in operation. Distances for other forms of energy, such as wave and tide, are not yet specified.

If public rights in the area are of paramount importance, say in a key fishing area, then consent could be denied even if a right to occupy the seabed has been obtained. A typical grid pattern for offshore wind turbines allows separation of 500 m or more and so free navigation through the area after construction is possible and in theory so is fishing. However, in conversation between the authors and fishers, many are reluctant to do so. They fear damage to either themselves, their vessels or their gear from underwater debris or inadvertent damage to the wind turbines or cables for which they would be blamed.

There is one more issue to address if legislation is to be changed. Todd points to the status of public rights as "incidents of property akin to servitudes".[14] As such, he says, they should count as "possessions" under the European Convention on Human Rights and the subsequent UK Human Rights Act 1998, and they can only be removed in the public interest.[15] Plant records that the government has "conceded . . . that the loss of access to fishing grounds by fishermen might in individual cases give rise to an entitlement to compensation".[16]

5   Emergence of the Blue Economy: challenges and lessons

There exists a global ambition to achieve economic growth and to create employment through expansion of the so-called Blue Economy. The oceans and seas contain ample resources with economic potential; both tangible (e.g. minerals) and intangible (e.g. transport routes and seascape). This potential is recognised in policies such as the Blue Growth Agenda pursued by the European Union.[17]

The challenges to these ambitions are twofold. First, the technologies that could recover these resources must do so at a commercially viable rate of return; second, there must be a recognised framework of jurisdiction, rights, ownership and management that would permit a peaceful and just regime of exploitation. This has to include protection for the environment and ecosystem services. It is probable that a clear incentive from the first (a commercial rate of return) is needed to encourage the second (the governance framework). The high value of offshore oil and gas reserves incentivised international agreement to allow their recovery. The potential value of seabed minerals encouraged debate, but volatile commodity prices have failed, so far, to incentivise technology and finalise agreement on rights and ownership. The world is left with a rather complex and piecemeal framework designed to meet the issues of the moment. Progress towards a recognised international order is painfully slow. However, transition from no ownership (with open access) or commons ownership (with rights for everyone) to enclosure and the introduction of private or state property rights is happening slowly. It is a necessary move, at least in affected areas, for the Blue Economy to flourish.

Recent EU-funded research into the Blue Economy has focused on eight sectors of existing or potential activity.[18] There are others, but these can be seen as main sectors influencing current policy. Four sectors are described as existing or established activities: offshore oil and gas, shipping, fisheries and tourism including recreation. The other four sectors are described as emerging, or Blue Growth, industries: aquaculture, marine and OE, seabed mining and biotechnologies. An immediate if superficial difference between these two groups is that the existing activities are driven by the market and, additionally in the case of fisheries, by the subsistence needs of some coastal communities. They need no government incentive, and the only state involvement is to regulate in the public interest supported by taxation for public benefit. The latter Blue Growth sectors are not supported by the market or not yet supported by the market. They are seen, for one reason or another, as desirable activities to be supported in their fledgling years by public policy, including financial subsidy. There is an old adage that governments do not pick winners, and the policy carries significant risk to taxpayers.

5.1 Existing industries

The offshore oil and gas industry is a relative newcomer to the marine scene but is already in quite rapid decline, although it will remain a feature for a number of decades to come.[19] Of the existing sectors, it is the one that most represents the idea of the new heavy high-technology maritime industries that are fixed and spatially defined with permanent or semi-permanent platforms, devices and moorings. The resource in this case is of high value and non-renewable. Where accessible and recoverable, governments in the past have asserted ownership if discovered adjacent to their terrestrial territory. The Truman Declarations of 1945 made just such an assertion for the United States in relation to the Gulf of Mexico. Subsequently, international law and agreement have accepted the emerging custom and recognised national rights and ownership of such resources on the continental shelves of the countries concerned. The value of the resource and its geopolitical importance incentivised solutions to its exploitation in relation to other rights like navigation and fishing. The approach taken to the challenges of recovering offshore oil and gas has been pragmatic with a preparedness to buy off obstacles or grant powers that otherwise would not have been granted.

Fisheries and shipping activities have far different characteristics from off-shore oil operations. Their presence on the surface of the sea is always transient, and there is no finely detailed spatial definition except perhaps at the ports from which they operate. Both activities enjoy public rights enshrined in international and national law giving them freedom to roam, the freedom of the seas.

With increasing globalisation of trade, shipping has been on a strong trajectory of growth. Freight has doubled in the last 20 years and is forecast to double again in the next 20 years. Dry bulk carriers and container vessels have seen the strongest expansion both in number and size. Also growing strongly but with a lower level of impact by volume have been passenger cruise vessels. These vessels need space to operate, although away from the port they tend to ply established routes, or sea lanes, that need to be kept free. The retreat of the sea ice in the Arctic Ocean has opened the prospect of new routes, and year-round access is already achieved along the Russian Arctic coast. Port congestion does increasingly create fixed occupancy of space for shipping and closes these areas to other activities. A number of proposals have been made to construct so-called container hubs, allowing fast access to ultra-large container carriers (ULLCs) and redistribution to an increased number of small ports in smaller vessels.[20] These include floating hubs that are of a size equivalent to artificial islands with considerable repercussions for rights and ownership to the space. Fisheries need the same freedom to roam as shipping and the ability to chase or hunt shoals of fish. Section 2 of this chapter has explained the background to ownership of this resource and the rights-based approach to fisheries management. Tourism and recreation rely on freedom to roam and the attraction of access to open space for activities or peaceful reflection as the key to the attractiveness and success of the industry.

5.2 Blue Growth industries

The Blue Growth industries and more particularly renewable energy and aquaculture introduce new demands on ownership and rights and sea that have not before been experienced. For the purposes of this chapter, seabed mining and biotechnologies will be discounted. Seabed mining activities will mainly be restricted to areas under the high seas where the minerals of interest occur. Ownership of the resource is a problem, but use of the space will be transient. Biotechnology resource recovery raises questions of resource ownership in the high seas, but poses little threat in relation to the right to occupy space.

Renewable energy and aquaculture in some ways reflect the same situation faced by offshore oil and gas: the need for the exclusive right to occupy areas of marine space in all three dimensions (seabed, water column and sea surface). There are, however, significant differences that compound the problem and bring new challenges. Offshore oil, for all its profile and impact, carried with it relatively small footprints of space mainly a long way out to sea and away from the coast and other activities. Pipelines and terrestrial support infrastructure created concern, but most worries related to the risk of oil spill accidents are rare events, albeit with potentially devastating environmental consequences.[21]

Renewable energy and aquaculture operations look to occupy very large areas of sea in close proximity to the coast, well in sight of land and in the midst of busy coastal waters. The prospects for clashes of interest and 'material infringement' of the public rights to navigation and fishing are therefore very high but as yet untested. Fixed offshore wind has tested the limits to some extent. However, these wind farms have tended to be offshore, over 6 km from the coast and in very specific sites where water depth and other criteria allow development. A water depth of 50 m is normally taken as the working limit for fixed turbines. In these areas, solutions have been found and provisions made for exclusion zones. The development of floating wind turbines greatly increases the number of potential sites, but many of these will be even further offshore and away from the most congested areas. Aquaculture has looked, up to now, for space close to shore to obtain shelter. The development of new technologies such as submersible cages allows more offshore sites to be considered. Licensed aquaculture sites are traded on the open market creating quasi-private property sites.

OE, in the form of wave and tidal devices and arrays, appear likely to have a high impact on space and hence on ownership and rights. It is not yet certain because no large-scale arrays exist and most full-scale trials have involved individual devices: these include the Pelamis Wave Energy Converter in Orkney and Portugal; the Aquamarine Power Oyster (wave) in Orkney; and various tidal devices at the Orkney test site and the Marine Current Turbines (MCT) array in the Pentland Firth (Scotland). The early plans and scoping studies for the commercial deployment of Pelamis and Oyster detailed large arrays in coastal waters, close to shore, with complex networks of moorings, cables and pipelines. In the event, both the companies involved with these technologies went into administration, and it is not possible to say how the plans might have developed. Investors, including government investors, ceased to provide adequate funds as the trials took longer than expected and showed insufficient signs of success. Enough work was done to show that the interruption of free navigation and fisheries is inevitable in these already congested waters close to shore. Solutions to questions of ownership and rights for OE are not yet resolved and are crucial to the wave and tidal energy industries when pressure for progress is again applied. Research into wave energy has slowed considerably, but the search for commercially viable technologies continues. Tidal stream energy, although slower in deployment than anticipated, is making solid gains with commercial arrays under construction.

A final word in this section is reserved for the concept of multi-use platforms (MUPs). The European Union, among others, has advanced research into MUPs as a signature policy. The advantages anticipated are a saving of marine space, reduced impact on the environment and savings in operational costs through sharing of resources. The concept introduces a new role for OE as an enabling source of energy for other activities. The EU-funded MARIBE Project studied

MUP activities, including offshore container shipping hubs, aquaculture, desalination and seaweed farming on platforms powered by wave or wind energy.

6 Indigenous marine rights and OE projects

Internationally, aboriginal groups have sought a range of mechanisms to assert rights and ownership over terrestrial and marine resources. Case studies indicate that while rights are often recognized, outright ownership is limited. The following case studies highlight the efforts of OE exploration by aboriginal groups in Canada, Australia and New Zealand.

6.1 Canada

Canada's coastline is the longest in the world (243,792 km), and it is the shape and geography of this coastline upon which most settlement in Canada is patterned. Management, rights and ownership of this extensive coastline and sea are complex especially with regard to First Nation rights in Canada. Treaty and non-treaty agreements are the mechanisms by which First Nation and Inuit organizations assert their rights over terrestrial and marine resources. In Nunavut, the Inuit self-governing territory in northern Canada, claims related to terrestrial and marine resources were addressed in the Nunavut Settlement Agreement developed between the Government of Canada and the Inuit.[22] Marine rights were specifically addressed by 13 of the 14 articles in this agreement.[23] The Haida Nation, 90 km off the coast of British Columbia, addressed marine rights through a series of co-management agreements that led in part to the creation of Gwaii Haanas National Park and a National Marine Conservation area.[24] These agreements have not provided outright ownership over marine resources. Rather, the intent of these agreements was to protect traditional access to these resources for subsistence harvesting and spiritual purposes. While First Nation and Inuit peoples have asserted ownership of marine rights beyond traditional purposes, rights over marine resources related to energy development, for example, is still held exclusively by the Government of Canada and increasingly by the provinces.

Aboriginal development related to marine renewable energy is limited. Most renewable energy development by aboriginal communities is wind-based.[25] Yet extensive resources for OE exist adjacent to aboriginal communities. For example, of the estimated 42,000 MW of tidal energy resources in Canada, 30,000 MW are in Nunavut, centred in the region of the Hudson Strait. Despite the magnitude of these resources and recognition by the Canadian National Research Council (NRC) of its energy-generating potential, exploration of this potential by developers is limited.[26]

Haida Gwaii has received attention from developers that explored both wave and tidal energy development. Like communities in Nunavut, Haida Gwaii is remote and totally reliant on imported diesel fuel for generation of its electricity.[27] Given the rising cost of fossil fuels and challenges with transportation,

British Columbia Hydro, a Crown Corporation, released a Request for Expressions of Interest (RFEOI) to explore other options for electricity generation.[28] Three respondents to the RFEOI focused on tidal energy generation primarily in the Masset Sound region in northern Haida Gwaii where 37 MW of extractable power exists. Wave energy potential has been explored at five locations on its west coast.

Although First Nation and Inuit involvement in OE is limited, initiatives have been developed to support proprietary rights over traditional resources. In Nova Scotia, the Mi'kmaq Ecological Study (MEKS) explored traditional and contemporary uses of the Bay of Fundy, a site where tidal energy development is being actively explored.[29] The Nova Scotia Marine Renewable Energy Strategy also highlighted the importance of respecting aboriginal rights established by Section 35 of the Constitutional Act of 1982.[30] While these studies and strategies provide a recognition of aboriginal rights, they do not imply ownership of marine resources.

6.2 Australia

Between 1788 and 1992, the Australian legal system considered Australia to be *terra nullius*,[31] or 'land belonging to no one'.[32] This was reversed in 1992 by the landmark *Mabo* case,[33] which recognised that in certain circumstances a continued beneficial legal interest survived the acquisition of title by the Crown. This concept of Native title was then formalised by passage of the Native Title Act 1993 (NTA),[34] which clarified the legal position and attendant processes.

In the *Yarmirr* case,[35] the court laid the foundation for offshore Native title claims, holding that Native title could include a right to free access to the sea and seabed for a number of purposes but that such a right could not be exclusive because that would be inconsistent with both common law public rights to fish and navigate and the international law right of innocent passage. In the *Blue Mud Bay* case, the High Court ruled that claims under the NTA apply to tidal waters overlying Aboriginal land.[36]

As a federal jurisdiction without bespoke regulatory structures for marine energy, the Australian regulatory landscape is a complex array of different institutions and authorities at the local, state and federal levels.[37] No defined processes exist in relation to marine energy, so projects are consented under existing environment and planning law controls.[38] At the same time and despite advancements in the recognition of native title, the planning regime has been slow to react and adequately involve Indigenous people or consider their rights.[39] In relation to fossil fuels, onshore projects require negotiation with Native titleholders, but offshore projects do not.[40] The NTA nonetheless requires that compensation be made for infringement of offshore rights.[41]

There are 18 developers actively investigating marine energy projects.[42] Ten projects are in the pipeline, representing 25 MW of wave energy and 535 MW of tidal.[43] Many projects are in the early stages of planning. While few projects to date have necessitated Indigenous engagement, one company, Tenax, is consulting Indigenous communities in the development of two tidal energy projects. The first, near Darwin in the Northern Territory, will not itself be situated within an Indigenous community but nonetheless affects such communities and requires consultation, particularly as the region is on the border of two different Land Councils.[44] The second, in the Kimberley region, involves close collaboration with the Indigenous community: a 'sea people' numbering 500, with strong maritime traditions, an in-depth knowledge of the climatic nature of the region and an appreciation of the potential of the tidal resource.[45] The community has shown keen interest in the project, and there has been high buy-in, particularly due to the potential for the project to support the economic life of the community.

The company notes that, despite minimal legal obligations, there is an elevated interest in Indigenous engagement.[46] Indeed, they see cultural issues as potentially being more important to their projects than strict legal obligations. In this regard, they note the need for slow, methodical and long-term consultation and for ensuring that projects are a win-win for communities.

6.3 New Zealand

British settlers signed the Treaty of Waitangi (1840) with the Māori, recognising Māori ownership of their lands. The Treaty has consistently been the subject of considerable debate, and, since the establishment of the Waitangi Tribunal (1975), Treaty claims and settlements have been a significant feature of NZ's political landscape. The common law recognises Maori customary title; the practice of the government was that Māori title extended over the whole country and had to be extinguished before it could be granted to new settlers.[47] Māori land tenure now comprises complex overlapping rights, with a number of methods of acquiring title.

Sea tenure has been a particularly controversial issue. After the Court of Appeal overturned the long-held assumption that the Crown owned the foreshore and seabed in 2003,[48] the Government passed highly contentious legislation vesting absolute ownership of the foreshore and seabed in the Crown and guaranteeing public rights of access.[49] A wave of protest ensued, and the UN Committee on the Elimination of Racial Discrimination noted that the legislation discriminated against the Maori by extinguishing the possibility of customary title over the foreshore and seabed without providing a means of redress.[50] The *Marine and Coastal Area (Takutai Moana) Act 2011* (MACA) repealed the legislation and instituted a 'no-ownership' principle,[51] restoring the possibility of customary marine title.[52] Such title is not exclusive but is subject to public rights and other specific interests.

There are no specific governance structures in place for MRE projects, and they are consented under a range of existing legislative instruments.[53] The *Resource Management Act 1991* (RMA) provides the core environment and planning framework and requires a permit to be sought for marine energy projects.

Consultation with stakeholders, including Maori groups, is not mandated by the RMA, though the Government nonetheless recommends that consultation takes place.[54] A developer can hold a resource consent but no formal lease over the seabed itself; the MACA treats any installation as merely personal property, not signifying any further legal interest.[55]

As many as 20 marine energy projects have been proposed over the last decade.[56] Of five projects receiving Government funding,[57] only three remain operative, and two other privately funded projects have become dormant or have been abandoned. One project, the Crest Energy tidal power project, underwent a lengthy approval process in NZ's Environment Court.[58]

The treatment of Maori cultural issues was a defining feature of the Crest case. The Te Uri o Hau Settlement Trust (TUOH) argued that granting consent would give Crest a *de facto* property right, thereby prejudicing its own ongoing High Court claim to customary ownership.[59] The significance of the area to TUOH had previously been recognised,[60] and the Court said the issues were "strongly in the mix".[61] Nonetheless, it noted that Crest had "adequately and appropriately addressed" the issues[62] through "extensive, considerable and meaningful" consultation and modification of its proposal.[63] The Court said that Crest had "essentially stepped into the middle of this situation".[64]

As a result, the project was consented in spite of TUOH's concerns. TUOH responded by issuing an *aukati*, a traditional ban that forbids Crest from entering the area, claiming that the RMA recognises its authority to do so.[65] Crest in turn cited its legal rights conferred by the RMA consents and stated its intention to ignore the ban, leading TUOH and the wider community to threaten direct action if Crest proceeded, in particular by organising a flotilla to intercept Crest's contractors.[66] Local politicians seized the moment to voice their concerns, with both a local MP and a candidate strongly expressing their opposition.[67]

7 The future of marine rights and ownership

To read of the ambition for the Blue Economy leads to a logical conclusion of a significant increase in private or state property rights granting exclusive use to areas currently guaranteed free for navigation and fishing as of public right. Such a move requires a reallocation of ownership and rights with old rights curtailed and new rights awarded. This is politically difficult and the sort of question where progress is made only when the pressure of development is so great that decisions have to be made. So it has been with offshore oil and gas, aquaculture and fixed offshore wind. Piecemeal and sector-specific arrangements have allowed these industries to develop so far.

Oil and gas is in decline, but aquaculture and wind power are forecast for strong expansion that may test the pragmatic and partial solutions to the limit. The introduction of floating offshore wind devices introduces new challenges to ownership and rights regimes. Universal marine spatial planning (MSP) is a stated objective at the international and national levels, but the future of the

Blue Economy industries is so uncertain with regard to technology, design and extent that most marine planning remains vague and anxious to avoid closing out development options. Hard and fast marine planning is restricted to sea areas like the Belgian part of the North Sea that is small and already so developed that zoning is self-evident. Policy-based planning, setting base criteria, or soft-zoning (or locational guidance) are the alternatives pursued.

Wave and tidal energy arrays have the potential to exhibit particularly difficult characteristics in relation to ownership and rights. The occupation of very large areas of sea space close to shore is bound to interfere with existing activities and raise questions of priority beyond those experienced so far with other recently developed industries, such as aquaculture and offshore wind. Solutions will be found, but it is likely to require real development pressure to create the will needed to find them. Compensation to fishers and those displaced may have to play a role.[68]

Notes

1 Scott, A. D. (2000) Introducing Property in Fisheries Management. In: Shotton, R. (ed.) *Use of Property Rights in Fisheries Management: FAO Fisheries Technical Paper 404/2*. Rome: Food and Agriculture Organization of the United Nations, pp. 1–13.
2 Hardin, G. (1968) The Tragedy of the Commons. *Science*, 162(3859), pp. 1243–1248; Grafton, R. Q., et al. (2006) Incentive-based Approaches to Sustainable Fishers. *Canadian Journal of Fisheries and Aquatic Sciences*, 63(3), pp. 699–710.
3 Scott (2000) n. 1; Arnason, R. (2005) Property Rights in Fisheries: Iceland's Experience with ITQs. *Reviews in Fish Biology and Fisheries*, 15(3), pp. 243–264.
4 Christy, F. T. (1982) *Territorial Use Rights in Marine Fisheries: Definitions and Conditions: FAO Fisheries Technical Paper 227*. Rome: Food and Agriculture Organization of the United Nations.
5 Grafton, R. Q., Squires, D., and Kirkley, J. E. (1996) Private Property Rights and Crises in World Fisheries: Turning the Tide? *Contemporary Economic Policy*, 14(4), pp. 90–99.
6 Bromley, D. W. (2016) Rights-based Fisheries and Contested Claims of Ownership: Some Necessary Clarifications. *Marine Policy*, 72, pp. 231–236.
7 Russ, G. R., and Zeller, D. C. (2003) From *mare liberum* to *mare reservarum*. *Marine Policy*, 27(1), pp. 75–78.
8 Grafton, R. Q., Squires, D., and Kirkley, J. E. (1996) n. 5.
9 Grant, G., et al. (2010) Cumulative Effects, Creeping Enclosure, and the Marine Commons of New Jersey. *International Journal of the Commons*, 4(1), pp. 367–389.
10 Cardwell, E., and Gear, R. (2013) Transferable Quotas, Efficiency and Crew Ownership in Whalsay, Shetland. *Marine Policy*, 40, pp. 160–166.
11 Newell, R. G., Sanchirico, J. N., and Kerr, S. (2005) Fishing Quota Markets. *Journal of Environmental Economics and Management*, 49(3), pp. 437–462; Gibbs, M. (2010) Why ITQs on Target Species Are Inefficient at Achieving Ecosystem Based Fisheries Management Outcomes. *Marine Policy*, 34(3), pp. 708–709.
12 Hansard (1973) *Zetland County Council Bill, Second Reading*. London: UK Parliament.
13 Johnson, K., Kerr, S., and Side, J. (2013) Marine Renewables and Coastal Communities – Experiences from the Offshore Oil Industry in the 1970s and Their Relevance to Marine Renewables in the 2010s. *Marine Policy*, 38, pp. 491–499.
14 Todd, P. (2012) Marine Renewable Energy and Public rights. *Marine Policy*, 36(3), pp. 667–672.
15 Ibid.

16 Plant, G. (2004) Offshore Renewable Energy Development and the Energy Bill. *OGEL*, 2(2004). Available at: www.ogel.org/article.asp?key=1236

17 EU (2014) *Communication from the Commission: Innovation in the Blue Economy: Realising the Potential of Our Seas and Oceans for Jobs and Growth* – COM(2014) 254/2 (13 May 2014).

18 Maribe (2016) *Deliverables*. Available at: https://maribe.eu

19 OECD (2016) *The Ocean Economy in 2030*. Paris: OECD Publishing.

20 Maribe (2016) n. 19.

21 Johnson, K., Kerr, S., and Side, J. (2013) n. 13.

22 Government of Canada (n.d.) *Government of Canada's Approach to Implementation of the Inherent Right and the Negotiation of Aboriginal Self-Government*. Quebec: Department of Indian and Northern Affairs.

23 Government of Canada (n.d.) n. 22.

24 Government of Canada (2010) *Gwaii Haanas National Marine Conservation Area Reserve and Haida Heritage Site Interim Management Plan and Zoning Plan May 2010*. Parks Canada.

25 Campbell, D. (2011) *More Than Wind: Evaluating Renewable Energy Opportunities for First Nations in Nova Scotia and New Brunswick*. The Atlantic Aboriginal Economic Development Integrated Research Program, AAEDIRP. April 2011.

26 Gillies, B. (1995) *The Nunavut Final Agreement and Marine Management in the North*. Canada: Nunavut Tunngavik Inc. Published by the Canadian Arctic Resources Committee, Vol. 23.

27 Blachfield, G., et al. (2008) Tidal Stream Resource Assessment for Masset Sound, Haida Gwaii. *Proceedings of the Institution of Mechanical Engineers, Part A: Journal of Power and Energy*, 222(5), pp. 485–492; Boronowski, S. (2012) *Integration of Wave and Tidal Power into the Haida Gwaii Electrical Grid*. Unpublished master's thesis, University of Victoria.

28 BC Hydro (2012) *Haida Gwaii Request for Expressions of Interest*, December 2012.

29 Membertou Geomatics Consultants (2009) *Mi'kmaq Phase 1 – Bay of Fundy, Nova Scotia Including the Fundy Tidal Energy Demonstration Project Site: Mi'kmaq Ecological Knowledge Study*. Available at: www.oera.ca/wp-content/uploads/2013/04/MEKS-Phase-I-Final-Report.pdf

30 Government of Nova Scotia (2012) *Marine Renewable Energy Strategy*. Department of Energy. Available at: www.gov.ns.ca/energy/resources/publications/nova-scotia-marine-renewable-energy-strategy-may- 2012.pdf

31 A Latin expression of Roman law origins, used in international law to describe territory not yet subject to the sovereignty of any state.

32 *Milirrpum v Nabalco Pty Ltd*. vol. 17 FLR 141. (1971).

33 *Mabo v Queensland* [No. 2]. vol. 175 CLR 1. (1992).

34 Subsequently amended by three Native Title Amendment Acts in 1998, 2007 and 2009.

35 *Commonwealth v Yarmirr*. vol. 168 ALR 42. (2000).

36 *Northern Territory of Australia v Arnhem Land Aboriginal Land Trust*. vol. HCA. 29. (2008). Coastal Aboriginal land is usually granted to the low water mark including the intertidal zone.

37 Wright, G. (2011) *Marine Renewable Energy: An Overview of Applicable Australian Legislation and Regulatory Bodies*. Available at: www.glenwright.net/files/Marine renewable energy legislation, Australia.pdf

38 Wright, G. (2011) Marine Energy in Australia and New Zealand: Regulatory Barriers and Policy Measures. In: *All-Energy Australia*, Melbourne 2011.

39 Wensing, E. (2007) Aboriginal and Torres Strait Islander Australians. In: Thompson, S. (ed.) *Planning Australia: An Overview of Urban and Regional Planning*. Melbourne: Cambridge University Press, pp. 254–275; Planning Institute of Australia Indigenous

Planning Working Group (2010) *Improving Planners' Understanding of Aboriginal and Torres Strait Islander Australians*. Kingston: Planning Institute of Australia.

40 Hewitt, A. (n.d.) *The Right to Negotiate: Where and When Does It Apply?* The University of Adelaide Abstract.

41 *Native Title Act 1993* s. 24NA (6).

42 Clean Energy Council. Available at: www.cleanenergycouncil.org.au/

43 Ernst & Young (2013) *Rising Tide: Global Trends in the Emerging Ocean Energy Market*. EYGM Ltd.

44 Major A. (2014) Pers. comm.

45 Ibid.

46 Ibid.

47 The Māori lost land to the Crown and private owners through a variety of methods, though usually through purchase. The dominant acquirer, by purchase or otherwise, was the Crown, even after the first Native Lands Acts (1862 and 1865) set up the Native Land Court to investigate Māori land titles. See Boast, R. (2012) Story: Te tango whenua – Māori land alienation. In: *Te Ara; The Encyclopaedia of New Zealand*. Government of New Zealand. Available at: https://teara.govt.nz/en/te-tango-whenua-maori-land-alienation

48 *Attorney-General v Ngati Apa*. vol. 3 NZLR 643. (2003).

49 Foreshore and Seabed Act 2004. See Charters, C., and Erueti, A. (2007) *Māori Property Rights and the Foreshore and Seabed: The Last Frontier*. Wellington: Victoria University Press.

50 Committee on the Elimination of Racial Discrimination. New Zealand Foreshore and Seabed Act 2004. Decision 1 (66) 2005. Charters, C., and Erueti, A. (2005) Report from the Inside: the CERD Committee's Review of the Foreshore and Seabed Act 2004. *Victoria University of Wellington Law Review*, 36(129), pp. 257–290.

51 Marine and Coastal Area (Takutai Moana) Act. 2011 s 11(2).

52 Ibid. s. 6(1).

53 Wright, G. and Leary, D. (2011) Marine Energy. *New Zealand Law Journal*, (August), pp. 227–230.

54 New Zealand Ministry for the Environment (n. d.) *An Everyday Guide to the Resource Management Act*. Available at: www.mfe.govt.nz/publications/rma/everyday/consent-apply/

55 ss. 18(2)(a) & (b).

56 AWETA. Marine Energy – NZ projects. Aotearoa Wave and Tidal Energy Association. Available at: www.awatea.org.nz/marine-energy/S

57 Through the Marine Energy Deployment Fund (2008–2011).

58 Re: Crest Energy Kaipara Limited. vol. NZEnvC A13. 2009. See Wright, G. (2011) A Tidal Power Project. *New Zealand Law Journal*, (September), pp. 260–261.

59 Paras 170–171. TUOH also expressed the same concerns regarding navigation, fishing etc. as the Court.

60 Te Uri of Hau Claims Settlement Act 2002.

61 Re: Crest Energy (2009) n. 59, para 181.

62 Ibid. para 214.

63 Ibid. para 198.

64 Ibid. para 175.

65 TUOH claimed that s. 6(e) of the RMA, which acknowledges the relationship of Maori with ancestral lands and water, extends to recognition of traditional bans. See Phillips, S. (2011) *Cultural Ban on Tidal Turbine Proposal, Suite*.

66 Downey, R. (2011) *Harbour Turbine Project Draws IWI Fire, Stuff*. Available at: www.stuff.co.nz/business/industries/4876726/Harbour-turbine-project-draws-iwi-fire

67 Ibid.

68 Todd, P. (2012) n. 15.

Part III

Project consenting and regulation

Chapter 7

Consenting ocean energy projects

Issues, challenges and opportunities

Anne Marie O'Hagan and Glen Wright

1 Introduction

Most jurisdictions have not yet implemented specific consenting processes for ocean energy (OE), and instead rely on a patchwork of existing instruments to consent various aspects of OE projects. The inefficient and ad hoc consenting frameworks that result have been consistently recognised as a major barrier to the progress of OE. Policymakers are, however, beginning to modernise regulatory frameworks in order to better facilitate and manage OE projects. In particular, some jurisdictions have started streamlining their regulatory frameworks, for example through the establishment of a one-stop shop (OSS) for consenting[1] and the development of targeted legislation and regulatory processes.

In this chapter, we first highlight the various key elements of consenting processes for OE projects before discussing in more detail the issues and challenges that unreformed consenting processes present. In Sections 4 and 5, we consider the opportunities and options for improving consenting processes. Section 6 presents a more detailed case study of the UK, a jurisdiction that has made considerable efforts to reform its regulatory framework to better support the sustainable development of OE projects (the following chapter provides shorter summaries of consenting processes in selected other jurisdictions). Finally, in Section 7 we offer some closing remarks.

2 Elements of consenting for ocean energy projects

A range of issues may be relevant in the permitting process. These include environmental impact assessment (EIA) and strategic environmental assessment (SEA); the rules regarding the use of ocean space, including compliance with any marine (MSP) processes; mediation of conflicts of interest, such as with fishers and surfers; and competing uses, like fossil fuel extraction. There are also likely to be rules regarding the extraction of energy itself or the extraction of water, requirements regarding construction, deployment and decommissioning, and established frameworks for the onshore components.

2.1 Occupation of marine space

OE developers will require sea space in order to develop projects and will need to obtain some form of rights for exclusive occupation as the basis for securely occupying the marine space, using the marine resources, and deploying OE devices. This may be in the form of exclusive rights over both the resources and the physical marine space in which those resources are contained.[2] Even if such rights are not explicitly sought, the needs and modalities of many OE technologies will nonetheless exclude other users, establishing a rights-like occupation of the marine space.[3] The OE industry therefore has the potential to present a major challenge to traditional conceptions of rights and "may play an important role in the redistribution of ownership rights in the marine environment" (see Chapter 7).[4] Furthermore, the physical space available for the deployment of OE devices is limited, with interest in developing projects coalescing on discrete areas of high suitability. This convergence of new industrial occupation and competition requires governments and policymakers to take difficult decisions regarding the allocation of rights to marine space and resources.

2.2 Exploitation of marine resources

As the oceans become increasingly industrialised, it is likely that there will be a concomitant increase in competition for access to resources. OE is particularly susceptible to such competition as the most viable resources are concentrated in specific locations. Consenting regimes will be at the front line in addressing a range of questions relating to resource management. Who has rights to access and use the resource? How are such rights determined? Who is in charge of making such decisions? Who is granted access to the resource and under what conditions? These questions are not abstractions but important issues that will affect the development of the OE industry, its impact on other marine users, and the value ultimately delivered to the public and to the environment.

2.3 Environmental impacts

Consenting processes are inextricably linked with environmental regulations. Almost all countries have EIA regulations in place, and the EIA process is often linked to consenting, either by being a precondition to or by being part of an integrated regulatory process. In EU countries, for example, there are a number of overarching legal instruments that implicate environmental considerations in consenting, such as the Habitats Directive,[5] the EIA Directives[6] and Strategic Environmental Assessment (SEA) Directive, the latter being relevant to public plans and programmes that cover the energy as well as other sectors.[7]

2.4 Terrestrial planning

While OE is a marine technology, devices will require auxiliary infrastructure on land. Cables and pipelines will make landfall on the coast, whilst substations and supporting infrastructure are necessary to transport the energy generated. This presents a challenge for consenting systems, which generally deal with terrestrial and marine activities on a separate basis according to separate legal frameworks and norms. Efficient and effective consenting regimes for projects that cross the land-sea divide will need to consider how the two separate systems might be integrated.

2.4.1 Electrical connection

OE requires infrastructure to connect projects to the electricity network. At the project level, grid connections tend to be governed by distinct legal and administrative processes and obtained from the national energy ministry or agency under an electricity act or equivalent. In practice, a range of other conditions must be complied with (including planning permission, EIA, a connection offer from the relevant operator and a power purchase agreement (PPA)[8] before consent to construct or generate electricity can be acquired. In addition, separate permissions are often required to lay cables. Beyond site-level considerations, the provision of electricity infrastructure presents strategic challenges in general,[9] and there is, therefore, a need to strategically consider long-term grid infrastructure needs and planning along with site-level connections during project development and consenting.

The variety of permits, licenses and permissions involved in consenting the electrical elements of a project are complex, time-consuming and often difficult to integrate into a single project license. Only a handful of jurisdictions operate a single permit system for all renewable energy projects.[10] This issue has been repeatedly raised in the EU, where the Directive on Renewable Energy requires streamlining of these processes.[11]

2.4.2 Insurance and liability

There are important practical questions regarding insurance and liability, which are increasingly relevant as projects develop. Indeed, as early as 1976, commentators expressed concern over the "thorny question of who bears the burden of responsibility for damage to or by [OE] devices".[12] The unfortunate case of Australian device developer Oceanlinx highlights these issues. Its flagship device sunk during transportation, resulting in a lengthy insurance dispute,[13] the company went into receivership, leaving it unclear who would take responsibility for the decommissioning of a second device that had been abandoned some four years earlier.[14]

2.4.3 Decommissioning

An improperly decommissioned device can become a significant environmental burden, and the failure to decommission in itself may be considered dumping.[15] The UN Convention on the Law of the Sea (UNCLOS) requires that any installations or structures that are abandoned or disused must be removed to ensure safety of navigation, taking into account any relevant international standards and with due regard to fishing activities, protection of the marine environment and the rights and duties of other states.[16] According to the relevant International Maritime Organization (IMO) Guidelines and Standards,[17] any installations or structures that are abandoned or disused in the exclusive economic zone (EEZ) must be removed to ensure safety of navigation in accordance with any accepted international standards,[18] except where non-removal or partial removal is consistent with the IMO Guidelines and Standards. Decisions should be made on the basis of a case-by-case evaluation, taking into account the following matters: potential effect on safety of navigation or other uses; deterioration of material and future effects; potential effect on the marine environment including living resources;[19] risk of shift from position; costs, technical feasibility and risk of injury to personnel; and determination of new use or other reasonable justification.[20] A coastal state may decide that the installation or structure does not need to be removed or that a decommissioned device can remain in situ if it would serve a new use such as an enhancement of a living resource.[21]

2.4.4 Health and safety

There are legal considerations surrounding health and safety regarding any industrial development, though such considerations are particularly pronounced in the offshore environment, which presents a range of potential risks to those tasked with working on marine projects. A range of international legal instruments contain specific provisions on various aspects of health and safety.[22] The provisions of these conventions are transposed into national legislation, and this will specify the responsible authority and necessary requirements and permits. Coastal states may have supplementary health and safety legislation seeking to address common activities and hazards, but in most cases this has not specifically been designed with the offshore environment or OE development in mind. To date the approach to health and safety for OE projects has been to extend the provisions of existing legislation, although some jurisdictions are adopting voluntary codes of conduct in an effort to ensure that OE is subject to the same or similar practices as land-based workplaces.[23]

2.4.5 Navigation

Responsibility for navigational safety in the marine environment rests with the IMO and national governments. A navigational impact assessment is usually conducted as part of project planning or for EIA purposes. This identifies where problems may arise as well as determining appropriate mitigation measures, such as marking the site with navigation aids.[24]

In the territorial sea, UNCLOS provides that coastal states have a right to adopt laws and regulations for the safety of navigation and, in particular, may adopt sea lanes, routeing systems and traffic separation schemes in order to ensure the safety of vessels and avoid collision.[25] As OE installations become more prevalent, it is likely that they will be increasingly sited within or near existing shipping lanes. The UK's Wave Hub,[26] for example, is situated close to a busy shipping area. During development of that project, radar and automatic identification system (AIS) surveys were carried out to better understand the shipping routes used in the area,[27] and the UK Government, in conjunction with the IMO, extended an existing Traffic Separation Scheme to ensure that traffic is kept away from the area.[28]

Coastal states also have the discretion to establish reasonable safety zones around structures and installations and to adopt appropriate measures therein in the interests of safety both of navigation and of the structures themselves. The IMO recommends that governments consider the establishment of such safety zones around offshore installations, as well as the establishment and charting of fairways or routeing systems through exploration areas.[29] OE devices are generally less visible than offshore wind turbines and will have mooring cables and anchor points that will need to be reflected in the dimensions of any operational safety zone.

3 Issues, challenges and opportunities

In most jurisdictions, the regulatory framework for OE projects is based on a range of legal instruments that are not tailored to the marine environment or to OE. The default position in most jurisdictions is therefore that the "legal planning framework has not been fully developed yet, forcing the authorities to create such a legal framework during the development of the project".[30] The result is that consenting frameworks are often a patchwork of procedures and permits (see Chapter 9) and are "the major threat to efficient implementation of this renewable energy source".[31]

In some jurisdictions, authorities have begun to mould this array of instruments into a more rational and consistent framework for consenting OE projects, while the most advanced have begun to more fully reform consenting frameworks with the marine or OE projects in mind. Nonetheless, considerable regulatory uncertainty remains in many jurisdictions, and consenting processes are generally poor because:[32]

- Regulators frequently rely on ad hoc consenting processes that are liable to change from one project to the next;
- Information regarding the relevant process is often difficult to obtain;
- There is often no clearly identifiable licensing authority;
- Statutorily defined timelines (whereby a regulatory authority must make its decision and communicate it to a developer within a specific time) are uncommon;
- Regulators often lack the requisite knowledge regarding the technology or legal context;

- The process can be unduly onerous; and
- Small-scale test deployments often face the full gamut of existing regulatory processes.[33]

Such permitting processes are not fit for purpose and give proponents little continuity or certainty. In many cases, the permitting process for a project, particularly one at large- or commercial-scale, can take several years, causing substantial delay and producing an undesirable level of uncertainty relative to the large level of investment required.[34]

From a regulator's perspective, burdensome and complex administrative procedures can prevent translation of high-level policy measures, such as government commitments to renewable energy deployment, into concrete action, such as assisting developers get approvals. Regulators are generally risk averse and are unlikely to assume responsibility for permitting projects they perceive as risky or to give priority to new technologies.[35]

In addition, different legal frameworks exist within individual countries at the regional or local level, e.g. in the UK, while consenting and planning rules vary significantly across jurisdictions, such as within the EU. Even once appropriate regulations have been developed, a lack of harmonisation among different jurisdictions, whether intra- or interstate or international, could hinder development of the industry.

3.1 Legal basis

While the UNCLOS provides a solid legal foundation for states to exploit OE resources within their national jurisdiction,[36] a large number of existing domestic legal instruments have the potential to impact the development of OE. As OE incorporates a range of regulated activities covered by differing pieces of legislation and administrative authorities, it is somewhat inevitable that the consenting system is convoluted and ad hoc in many places. The assortment of consents, licenses, leases, permits, authorisations and permissions involved in consenting are complex and time-consuming individually and consequently are often difficult to integrate into a single project license.

In the absence of specific legislation designed to rationalise the system and provide an institution with an appropriate mandate, there is unlikely to be an appropriate legal basis for an existing institution or agency to undertake this task. Evidence from jurisdictions that are making progress in streamlining consenting processes suggests that this can be challenging, often requiring substantial legal amendment as well as high levels of investment and political commitment,[37] which can sometimes be scarce for new and developing sectors. The implementation of a so-called one-stop shop for consenting has emerged as the front-runner response to these issues and is discussed in more detail later in the chapter. The OSS approach is generally viewed favourably by developers, but, depending on the model adopted, they can also have significant resource implications and may require formal legal amendments.

In the EU context, the Renewable Energy Directive explicitly provides that "administrative procedures are streamlined and expedited at the appropriate administrative level" and that such procedures be "clearly coordinated and defined, with transparent timetables for determining planning and building applications".[38] The Directive also envisages the application of "simplified and less burdensome authorisation procedures" to smaller projects and for decentralised devices producing renewable energy.[39] In theory, this should provide a legal basis for Member States and competent national authorities to develop more appropriate regulation, but progress has been slow. A majority of Member States recognise the need for more improvements in their respective administrative systems as they apply to renewable energy,[40] but to date only Denmark, Italy and the Netherlands have a single permit system for all renewable energy projects.[41]

Overall, the legal basis for consenting of OE is well established, but the procedures involved in its administration remain multifaceted and difficult. This could change as the number of operational OE deployments increase and problematic issues become increasingly apparent. Whilst progress has been made in some domestic legal systems, an interconnected and consistent framework is an aspiration for many countries and developers alike.

3.2 Environmental impacts and assessment

EIA has become almost a prerequisite for consent in most countries. The integrity and protection of the marine environment is of paramount concern to competent authorities that sanction development in that area and can also trigger very strong legal obligations. While there is concern that regulatory and governance processes will be relaxed so as to pursue and achieve a reduction in greenhouse gases, thereby causing "paradoxical harm" to local ecosystems,[42] OE can also result in positive environmental effects.

EIA is a site-specific assessment, so the parameters to be measured and included with the EIA itself are heavily influenced by the proposed location of the project. This is potentially problematic as there is little or no consistency in the methodologies applied to the study of specific parameters or questions posed in relation to them. This, in turn, limits the ability to draw inferences, identify trends and increase knowledge on the environmental effects of device deployments as different methodologies may produce different results and hinder comparison.[43] The ability of scientists to compare data and results across deployment sites is one way in which knowledge and expertise can be increased, hence advancing learning about these new technologies.

Environmental monitoring at test centres has increased data and information on the environmental effects of devices. This, coupled with international efforts to disseminate information on environmental effects,[44] is continuously adding evidence and generating knowledge; however, fundamental environmental research questions for wave and tidal energy remain. These are difficult to address at site level by developers and will require more concerted national and regional action, particularly in relation to migratory marine species, for example.

In the EU, an amended EIA Directive[45] aims to address the identified shortcomings in the EIA process, to reflect changing environmental and socio-economic priorities and challenges, and to align the EIA Directive and process with the principles of smart regulation. One of the changes that could impact consenting of OE projects is the provision enabling coordinated and/or joint procedures where a number of assessments have to be completed (i.e. EIA and "Appropriate Assessment" under the Birds and Habitats Directives). The effect of this is that a single assessment will be possible once the amended Directive applies. A revised screening procedure is introduced where Member States can set thresholds or criteria to decide when a project does not need to be screened or subject to an EIA. If a competent authority decides that an EIA is not needed, it must make this decision available to the public, along with the reasons why it is not required, as well as mitigation measures proposed by the developer to avoid or prevent significant adverse effects on the environment. The competent authority in the Member State is required to make its determination (on screening) within 90 days from the date on which the developer submitted the required information.[46] In an effort to ensure the quality and completeness of submitted EIAs, the developer must ensure that the EIA report is prepared by "competent experts", and the competent authority must ensure it has, or has access as necessary to, "sufficient expertise" to examine the EIA report.[47]

In terms of strategic environmental assessment (SEA), there appears to be limited application of SEA to OE development (see Chapter 10 for further discussion). This is a missed opportunity, as the documentation compiled for the SEA can become the first source of environmental information that developers consult during site selection and project planning stages. SEA can therefore inform site selection through the provision of relevant information, constraints mapping and the identification of low-sensitivity sites.

3.3. Public consultation and acceptance

Public acceptance is becoming more and more influential in the outcomes of decisions relating to the various consents needed to operate a marine or OE project. This has been colloquially termed a 'social license to operate'. Responsibility for engaging with the public and communities potentially impacted by a proposed development primarily falls to the developer or project proponent. Acceptance is neither automatic nor unconditional, and significant effort on how best to engage the public and allay any concerns they might have is required. Habitually, in past projects, the involvement of the public was almost entirely limited to the consultation phase of the EIA process or at the behest of the individual site developer if a specific issue arose. This routinely resulted in stakeholders expressing frustration with how they were involved in project planning, either by being consulted too late in the process or having limited influence on the decision made. A social license to operate is based on a multitude of principles including legitimacy, credibility and trust, which take time to foster and cannot be neatly slotted into a time-bound consenting process. A concerted effort to promote and understand OE technologies and their associated infrastructural requirements could help mitigate objection, thereby increasing the likelihood of social acceptance.

Generally marine renewable energy is viewed positively, but often there is a lack of familiarity and understanding of wave and tidal energy technologies in particular at local level given there have been relatively few deployments to date.[48]

There are also deeper questions in this context that go to the core of rights and ownership issues. Governments or their agencies are responsible for decision making in the marine space. Some domestic legal instruments provide that such decisions must be made in the public interest or for the common good, but it is not always acceptable that governments decide on the use or activity that can take place in what is considered by many to be a common resource. Through the implementation of MSP, questions regarding the allocation of marine space and its resources are likely to become more prominent, and, depending on the approach taken to MSP at the country level, plans developed at the regional or local level could be more reflective of the needs and desires of the public in that area for their adjoining marine space, including any future activities that might occur therein.

3.4 Marine spatial planning and new management approaches

The full implementation of MSP could provide an opportunity to improve consenting for many marine developments, including OE, through increasing transparency and providing greater certainty for both developers and their investors. MSP seeks to reflect environmental, social and economic interests in an inclusive way. As such, it has the potential to balance precaution and risk so as to provide flexibility but within a framework that is predicable, consistent and transparent to those involved.

The adaptive nature of the MSP process can react to changing circumstances, which is important for developing industrial sectors such as OE. It could promote coherence between terrestrial planning systems and those that operate in the marine, but this very much depends on how it will be implemented and enforced in each country. It is not yet clear how different competing activities will be accommodated in MSP or how decisions on 'trade-offs' will be made. Coexistence may be advocated as a preferred option, but this is not always technically or legally possible. Realities of health and safety concerns, insurance and liability could stymie coexistence before it even happens.

Apart from Scotland and, to a limited extent, maritime spatial plans applicable to certain waters around specific US states, many existing maritime spatial plans include only existing uses with little or no consideration or inclusion of new or innovative marine activities that could occur in future (see Chapter 6 for further discussion). As a relatively new approach to planning marine activities, it is vital that MSP and plans developed as part of that process promote the coexistence of relevant activities and uses and engage everyone in the process.

4 Options for reforming ocean energy consenting processes

Though poor regulatory frameworks for OE are common, they are nonetheless evolving in some jurisdictions. Offshore wind has triggered the formation of a more appropriate consenting system in many countries. To harness the available OE resource, investors and project developers look towards countries with low levels of 'regulatory risk', namely those with a stable, transparent and predictable regulatory system. This suggests that governments and their agencies need to be more proactive in how they deal with OE proposals. Consenting processes must aim to be both principled and practical, ensuring:

- **Economic efficiency.** As the rights being allocated are for the exploitation of a finite[49] and valuable resource, they should be allocated in a manner that ensures that the resource will be sustainably developed in a manner that maximises public benefit. This may involve a competitive allocation process (see later in the chapter).
- **Equity.** Ensuring that the resource is allocated equitably amongst legitimate proponents.
- **Sustainability.** It is crucial that regulatory processes ensure sustainable deployment of OE devices.
- **Financial return.** Where government opts to permit a private use of a shared/common marine space, it should ensure an appropriate level of financial return.
- **A simple and user-friendly consenting process.** The process should not add regulatory burden and risk and therefore time and cost to an OE project should be provided.

In addition to these points of principle, several important questions should be considered in the course of reforming OE consenting processes:[50]

1   How best to modify consenting processes so that they reflect the scale of development and the level of risk posed, in particular by imposing more permissive procedures for small-scale, time-limited deployments in areas of low environmental sensitivity.

2   How to facilitate the transition towards integration of the various competent regulatory bodies in consenting processes. In particular, whether it may be possible to extend the OSS approach further to also integrate grid connection, electricity licensing requirements and other incidental approvals. In this regard, it is important to facilitate and improve communication among regulatory bodies and to clarify their respective responsibilities regarding enforcement conditions.

3   Development of simple alternatives to OSS systems for jurisdictions where political will is insufficient to allow for more wide-ranging reforms.

4   Development and mainstreaming of effective consultation/participation processes in the marine context.

5   There are also deeper questions that go to the heart of the rights and ownership issues discussed in Chapter 7. Should decisions regarding marine resource allocation and OE development consents be made at a more local or regional level rather than at a national level? Where does the balance of power currently lie with regard to taking such decisions, and where should it lie?

In most jurisdictions, the processes for allocating site tenure and the processes for consenting a project are separate and involve different considerations. Allocation of tenure provides security that a project can use the desired resource (and therefore provides investment security for proponents and investment), while the relevant consenting processes and regulatory approvals determine whether a project can proceed in that location in accordance with existing environmental and other laws. We therefore discuss these two issues separately, though ideally the two aspects would be dealt with in an integrated manner in any eventual regulatory mechanism.

4.1 Allocation of site tenure

Several possible mechanisms or combination of mechanisms can be used to make decisions in relation to tenure allocation:

- **Developer-led permitting.** A first come/first served approach whereby developers apply for permits as and when they require them. Each project then proceeds through the relevant consenting processes in the order of application. This approach is the default position in jurisdictions that have not developed specific permitting processes for OE projects or other marine activities.

- **Qualitative assessment.** Allocation of tenure and approvals are based on a qualitative assessment of the proposals received. This will require the regulator to conduct a detailed assessment and comparison of the expected performance of a number of projects and allocating tenure on the basis of which project best fulfils set criteria.

- **Competitive processes.** Ranges from simple tenders to more sophisticated processes, such as online auctions and leasing rounds. Selection occurs with reference to bids that are assessed against key criteria (both qualitative, as previously discussed, and quantitative).

Developer-led permitting is perhaps the most problematic model. Such a process risks allocating development rights too early, shutting out project proponents that may be more suited to developing the resource but not yet in a position to make an application. Early movers may gain rights over the best sites, but the number of potentially interested proponents may be too small to enable efficient and effective competition. Developer-led permitting is not well suited to maximising the policy objectives previously identified: it is unlikely to be economically efficient as it provides no basis for assessing the strength of particular projects; it is unlikely to be equitable or provide a good financial return as it could be open to exploitation or could favour poorly prepared proposals; and while the environmental impact of these proposals would be covered by EIA legislation, projects put forward by well prepared and experienced proponents will likely be better optimised for positive environmental outcomes. Despite the inadequacies of this form of permitting, the model persists in jurisdictions where legal reform has not yet been forthcoming.

Qualitative assessment will place a greater burden on regulators, as an informed decision requires understanding and assessment of a range of factors, but the additional layer of assessment will better ensure that projects that proceed to the consenting phase are strategically selected to result in the best outcomes. This assessment could still take place on a first come/first served basis, or a window for applications could be opened to ensure a range of applications.

A competitive process seems most likely to provide a solid structure for the approvals process that can ensure that a range of suitable proposals are considered and that the principles previously mentioned are reflected. Nonetheless, there are also risks with a competitive process. Competition usually relies on having a sufficient number of well prepared competitors. In a fledgling sector such as OE, the competitive process may result in unrealistic or untested proposals. In addition, it may generate competitive sentiment between developers where it may be advantageous for them to work together to overcome common hurdles and develop a strong foundation for the sector, before they begin to compete for preferred sites.

Finally, there is the overarching issue of timing of tenure allocation, that is at what point should the tenure be awarded, relative to the other elements of the regulatory process, such as EIA and planning consents. In all jurisdictions, a range of legislative requirements are associated with an OE proposal. An important consideration is how to ensure that the tenure allocation process is structured and timed so as to create a logical regulatory sequence between allocation of tenure and regulatory approvals.

The basic choice is whether allocation of the rights to occupy the marine space should be allocated prior to the completion of all legislative requirements or afterwards. If tenure is allocated before all legislative requirements have been met, security of tenure will exist, but there is no guarantee that the project will meet all legislative requirements. This would leave the lease stranded. If tenure is only allocated after all legislative requirements are met, money and time may be spent obtaining the legislative requirements only for the preferred site to no longer be available, particularly as the industry grows and if there is no other undertaking or guarantee from government to reserve the preferred sites. This may also cause a 'lockout' effect: if a large number of developers apply for tenure options but do not make use of them immediately, this precludes the area from being used at all, potentially resulting in a large number of potential sites being effectively closed to other proponents.

4.2 Consenting processes

The main response to the problems with poor consenting processes has been to create an OSS for permitting applications. This essentially means concentrating the process in one regulatory body or authority. This authority can then liaise with the developer and work with the other relevant government departments and authorities to obtain the necessary consents. In this way, the developer has to face only one body rather than many, while the various licensing processes can be consolidated, coordinated and streamlined. In theory, this can reduce the burden on applicants by providing a single point of contact for developers, reducing the pressures on the licensing process by providing a more efficient use of available regulatory and human resources, enabling coordinated consultation with interested parties, and allowing for a more holistic assessment of projects.

The OSS approach has garnered much praise from developers and proponents of the OE industry.[51] The OSS idea has a history of being a preferred reform for developing offshore energy industries (having been implemented in the United States as part of early OTEC (ocean thermal energy conversion) efforts[52] and in relation to wind energy)[53] and is generally seen in a favourable light by developers.[54]

The development of a successful OSS takes political will. This is needed not only to ensure that the OSS is amply resourced but also to dismantle existing regulatory structures and overcome resistance from existing regulatory bodies that may perceive this streamlining as an unwelcome "centralisation and de-democratisation of decision-making".[55] The success or failure of an OSS and of OE policy in general is likely to turn on how highly OE is prioritised by government. In addition, despite the prevalence of discussion of OSS in the literature, recent research suggests that the actual level of implementation and likely benefits may have been overstated.[56] The implementation of the OSS approach might merely shift the burden from developers to administrators, thus resulting in the need for extra resources.[57] This suggests that any effort to implement OSS must be based on strong political will, adequate financing and support, and cooperation and collaboration among all parties.

There are also possible alternatives to the OSS approach. One option is to develop parallel consenting procedures that enable different issues to be evaluated simultaneously by existing regulators and expert groups. The coordination of such parallel processes in the absence of an OSS clearly creates additional administrative demands but may prove an attractive alternative for countries facing constraints in developing a single licensing authority.[58] A further alternative is the lead agency approach,[59] which could be regarded as a weak OSS approach. This is where an existing agency takes on the responsibility for coordination of parallel consenting procedures. The identification of a lead agency therefore eliminates the burden of dealing individually with a number of bodies in much the same way as an OSS; however, the lead agency will not be invested with additional powers or authority in the way that an OSS would be. The lead agency will also retain its existing statutory mandate, roles and functions, which will mean that it will not align as closely to the OE industry (or other relevant industries) as an OSS. Nonetheless, this approach could be effective where there is insufficient momentum for a devoted OSS or where there are state/federal issues.[60] Finally, an unexplored option, as yet, would be to establish an inter-agency task force or commission, whose members would be representatives of all potentially involved regulatory bodies. This could be used as a process for granting approvals in its own right or as a model for consultation and coordination among departments that could feed into departmental decision making.

5 Case study: the UK

In this section, we discuss the efforts made by the UK, in particular England and Scotland, to reform its consenting processes for OE projects.

5.1 Seabed tenure

The Crown Estate (TCE) manages the UK seabed and is responsible for allocating seabed tenure for OE projects in the UK (out to the 12-nm territorial sea limit). TCE is a statutory body tasked by Parliament with achieving particular goals in line with the principles for good consenting previously discussed,[61] and it has made an express commitment to work with all stakeholders to develop the OE industry. TCE has already provided leases for test and demonstration facilities,[62] as well as for other test and demonstration projects, and has held two commercial leasing rounds.[63] Its efforts to provide for OE are therefore not a reluctant regulatory response to an emerging problem, but a concerted effort to assist the industry and lead a balanced process to allocate resources.[64]

5.1.1 Pentland Firth and Orkney waters leasing round

The Pentland Firth and Orkney waters (PFOW) area was the first in the UK to be opened up for commercial-scale development of OE projects. This entailed a competitive leasing round for demonstration- and commercial-scale project sites, which received considerable interest from industry. TCE announced plans to hold a leasing competition in September 2008. Initial proposals for projects were invited from developers in November 2008, with an initial nominal target capacity of a total of 700 MW. Thirty-eight pre-qualified proponents were invited to apply for leases: 20 bidders applied with a total of 42 applications. These ranged from small development companies to multinational energy companies, with projects from 10-MW demonstration schemes to hundreds of megawatts.

TCE's process has been broadly successful in that it has attracted a range of developers to apply for leases. Pre-approval of bidders ensured that proposals were not received from ill prepared companies, and the process appears to have been a truly competitive one, thereby maximising adherence to the principles previously discussed. As the first process of its kind, it was always expected that it would generate institutional learning, which could be applied in future processes and other jurisdictions. Indeed, a number of important lessons can be learned.

While a competitive approach may work well for an established industry, it may have been disadvantageous in the present context for four reasons. Firstly, TCE's process appears to have limited site availability because, during the leasing process, TCE was not open to other applications. This led to the assertion that the process "seems too rigid to accommodate the fast moving nature of the growing marine energy industry" and that a number of developers who were developing projects outside of TCE's leasing rounds were disadvantaged by TCE's restriction on sites.[65] Likewise, another commentator said that TCE's approach should be one of "keeping constraints to a minimum and providing as much flexibility for deployment as possible".[66]

Secondly, TCE's process appears to have forced developers to compete at a time when they would have benefited more from cooperation. As one developer noted, "[A]t this early stage, collaboration may be more appropriate if we are to overcome the substantial common hurdles and risks".[67]

Thirdly, in aiming to kick-start commercial-scale development, TCE may have inadvertently "shut out" the demonstration-scale proposals that remain crucial to the industry's overall development. TCE developed its process in the context of offshore wind; however, OE is at a much earlier developmental phase, which means that a full-scale commercial leasing round may not have been the most effective option. The process also assumed that wave and tidal energy were at the same level of maturity – a view that is not shared across the sector. An OE project proponent must have a lease granted by TCE to commence a project, and there are two ways to obtain this: either apply for a demonstrator lease (at 10 MW or 20 devices) or bid in the competitive leasing process previously discussed. However, once an area is under competitive tender, TCE is unlikely to approve any demonstration leases within this area,[68] thereby effectively excluding demonstration projects from some of the best resources. This issue could easily be fixed by zoning an area for demonstration deployments within the larger leasing area.

The leasing process also highlighted the need for collaboration and cooperation between different bodies, particularly as MSP becomes more widespread.[69] A potential tension exists between MSP, leasing and consenting that must be managed. In the case of the PFOW leasing round, the TCE process proceeded in advance of MSP, and the local population were not included in the decision to move ahead with this. Ultimately, there was a strong public backlash against TCE process, which was perceived as giving a green light to projects and drawing lines on maps akin to zoning, before any consultation or EIA processes had taken place.

The British Wind Energy Association & Scottish Renewables note that uncertainty about future leasing rounds created difficulty for business planning, and asserted that:

> there is strong support within our wave and tidal membership for further leasing rounds to be open on a rolling basis, following SEA completion and market support, and for these to be set out in a planned programme so that industry can plan ahead.[70]

These difficulties could be described as inevitable teething problems, though at their root is the broader question of how a body like TCE can best balance the need to ensure that a sustainable industry emerges in the long term, while also meeting the shorter-term requirements of innovative developers keen to deploy their devices.

5.2 Permitting process

5.2.1 England

The Marine and Coastal Access Act 2009 (MCAA)[71] reformed marine licensing in England by consolidating and replacing some previous statutory controls. The Act provided for the creation of the Marine Management Organisation (MMO), now responsible for most marine licensing in English inshore and offshore waters and for Welsh (and Northern Ireland) offshore waters.[72] A marine license granted by the MMO is required for many activities involving a deposit or removal of a substance or object from the sea or a tidal river[73] and therefore incorporates OE projects.

There are two separate regimes for projects in English waters: one for renewable energy projects over 100-MW capacity, processed by the Planning Inspectorate,[74] and one for projects under 100-MW capacity, which are the responsibility of the MMO. However, while the MMO licenses marine elements of a project, other components of the project are licensed under different regulations, including:

- Section 36 consent (required under the Electricity Act 1989) to build and operate an energy generation site;
- Safety zone consent (required under Section 95 of the Energy Act 2004);
- European Protected Species license;

- The Department of Energy and Climate Change, responsible for project decommissioning under the Energy Act 2004; and
- The local Planning Authority, responsible for onshore planning.

5.2.2 Scotland

Consenting procedures for OE in Scotland are broadly similar to those of England and Wales but involve a distinct administrative system. Under the Marine (Scotland) Act 2010,[75] the Scottish Government, through Marine Scotland, is responsible for the new marine licensing system for activities carried out in the Scottish waters out to 12 nm.[76] Proponents will still require a Section 36 license under the Electricity Act, a European Protected Species license,[77] and decommissioning approval, each issued by separate bodies. Consent under the Town and Country Planning Act 1990 is also required.[78]

It is intended that the new system will enable consistent decision making about what activities are allowed to take place at sea. Through the process of marine licensing and the conditions placed on licenses, economically and socially beneficial activities are promoted while minimising adverse effects on the environment, human health and users of the sea.

In contrast with other parts of the UK, however, Marine Scotland has adopted an OSS system to provide a single contact for advice, enquiries and applications to simplify consenting and reduce the burden on applicants, regulators and other parties. The system is also intended to facilitate coordinated consultation with nature conservation bodies and other parties so as to promote interaction and more holistic assessment of proposed projects. Marine Scotland in its role as regulator is also tasked with ensuring compliance with the conditions of Section 36 license under the Electricity Act and the marine license.

5.2.3 A preliminary assessment

The consenting system in England continues to involve a number of authorities granting different licenses, and the resulting sequential process can still be fairly arduous for project proponents.[79] The MMO and TCE have agreed on a Memorandum of Understanding, which may in time lead to a more coordinated approach. In addition, the Senior Licensing Manager of the MMO has suggested some ways in which consenting procedures could evolve to become more effective, including:[80]

- Early engagement of key actors in order to streamline regulatory processes;
- All parties agreeing on regulators taking the lead to streamline consultation (this may be difficult where competence is spread across departments that may wish to retain their control over certain aspects of the process);

- Implementing MSP with the aim of increasing the likelihood of OE projects receiving consent (these still need to comply with relevant legislation); and
- Increasing regulator knowledge through a range of mechanisms.

Scotland's OSS has been well received and is generally perceived as providing developers with the greatest confidence in the regulatory process.[81] Some industry participants with real-life experience using the OSS have found that it is not always a truly integrated process[82] and that Marine Scotland must ensure that as the industry develops, it keeps a hold on the process. The Section Leader of the Licensing and Operations Team at Marine Scotland therefore also identifies early strategic engagement with all parties as key in improving consenting regimes. He additionally notes that there is a need to "stop re-inventing the wheel for every project",[83] suggesting that processes have not yet been standardised.

There is the question of whether the OSS concept can be replicated effectively in other jurisdictions, and there has so far been little discussion of how well OSS processes will fare once the industry has grown to its full potential, though at least one commentator has identified that OSS "may come under greater scrutiny as the sector continues to develop and larger, more contentious developments are proposed".[84] Large-scale developments are likely to put considerable strain on a single authority, and it is yet to be seen whether an OSS can cope with the range of issues that such developments will likely bring.

6 Conclusion

This chapter has highlighted that OE projects will have to go through a range of consenting processes before devices make it into the water. OE projects will need approvals for many aspects of their operation, including for the occupation of marine space, the exploitation of marine resources, the generation of environmental impacts, and decommissioning. To date, few jurisdictions have attempted to reform these regulatory frameworks in order to facilitate good governance of OE projects. As unwieldy consenting processes are commonly cited as one the major non-technical barriers to the development of OE, such reforms will likely be crucial, yet it is clear that this is not a simple task and that considerable political will is required. In the UK and Scotland, where substantial reforms have been implemented, there are early signs of success, and other jurisdictions seeking to develop an OE industry may be able to build on this experience.

Notes

1 O'Hagan, A. M. (2012) A Review of International Consenting Regimes for Marine Renewables: Are We Moving Towards Better Practice? *Fourth International Conference on Ocean Energy*, Dublin; Simas, T., et al. (2015) Review of Consenting Processes for Ocean Energy in Selected European Union Member States. *International Journal of Marine Energy*, 9, pp. 41–59.

2 Though not all jurisdictions necessarily allow for private or even State rights over the sea and seabed. See e.g. the discussion of New Zealand in the following chapter.

3 For example, a wave energy device sitting on the sea surface will require rights to occupy the surface but would likely also preclude use of the area below and around the devices.

4 Kerr, S., et al. (2014) Establishing an Agenda for Social Studies Research in Marine Renewable Energy. *Energy Policy*, 67, pp. 694–702.

5 Council Directive 92/43/EEC of 21 May 1992 on the conservation of natural habitats and of wild fauna and flora.

6 Council Directive 85/337/EEC of 27 June 1985 on the assessment of the effects of certain public and private projects on the environment as amended by Directives 97/11/EC, 2003/35/EC and 2009/31/EC; codified in Directive 2011/92/EU of the European Parliament and of the Council of 13 December 2011 on the assessment of the effects of certain public and private projects on the environment and subsequently amended in 2014 by Directive 2014/52/EU of the European Parliament and of the Council of 16 April 2014 amending Directive 2011/92/EU on the assessment of the effects of certain public and private projects on the environment. Note that under Annex I of the EIA Directive an EIA for wave and tidal projects is not mandatory. Rather they appear to fall under Annex II where the necessity to conduct an EIA is left to the discretion of the Member States. In practice, an EIA process will usually be in place.

7 See, e.g., Faber Maunsell and Metoc PLC. (2007) *Scottish Marine Renewables Strategic Environmental Assessment (SEA) Non-Technical Summary*. Edinburgh: The Scottish Executive. Available at: https://tethys.pnnl.gov/sites/default/files/publications/Scottish_Marine_Renewables_SEA_Summary.pdf; Doelle, M. (2009) The Role of Strategic Environmental Assessments (SEAs) in Energy Governance: A Case Study of Tidal Energy in Nova Scotia. *Journal of Energy and Natural Resources Law*, 27, pp. 112–144.

8 This is a contractual agreement between an electricity generator and a licensed supplier obliging the latter to purchase the output from a new renewable energy powered electricity generation plant.

9 Offshore energy resources are often located in remote coastal areas where there is only a weak distribution network available (if any), such that the costly grid reinforcements would likely raise project costs to a prohibitive level. There are often weak connections between States (e.g. EU Member States), power markets are generally inflexible and fragmented, and there is a lack of offshore electricity grids. See Van Hulle, F., et al. (2009) *Integrating Wind: Developing Europe's Power Market for the Large Scale Integration of Wind Power*. Brussels: European Wind Energy Association (EWEA). Available at: http://orbit.dtu.dk/files/3628704/Summary.pdf; Soerensen, H. C., and Korpås, M. (2010) Integration of Wave and Offshore Wind Energy in a European Offshore Grid. In: *Twentieth International Offshore and Polar Engineering Conference*. Beijing: International Society of Offshore & Polar Engineers, pp. 926–933.

10 Renewable energy progress report COM (2013) 175 final.

11 Directive 2009/28/EC provides that "administrative procedures are streamlined and expedited at the appropriate administrative level" and that such procedures are "clearly coordinated and defined, with transparent timetables for determining planning and building applications" (Article 13(1)(a) and (c)). The European Commission has recognized that progress in removing administrative barriers remains limited and slow, can raise the costs of renewable energy generally and will require further efforts if the 2020 targets are to be achieved (ibid.).

12 Knight, H. (1976) Legal, Political, and Environmental Aspects of Ocean Thermal Energy Conversion: A Report on an ASIL/ERDA Study. In: Kohl, J., *Energy from the Oceans: Fact or Fantasy*. Raleigh: Center for Marine and Coastal Studies, North Carolina State University, p. 45.

13 See Victoria Harbor Times (2014) *Oceanlinx Energy Generator at Carrickalinga Is Still Considered Unsafe*. Available at: www.victorharbortimes.com.au/story/2469811/oceanlinx-energy-generator-at-carrickalinga-is-still-considered-unsafe/

14 See Victoria Harbor *Times* (2014) *Carrickalinga Prohibited Zone Around Oceanlinx Wave Energy Device Attracts Big Fines*. Available at: www.victorharbortimes.com.au/story/2597306/carrickalinga-prohibited-zone-around-oceanlinx-wave-energy-device-attracts-big-fines/

15 Abandonment of human-made structures at sea potentially comes within the definition of "dumping" under relevant international law (1996 Protocol to the London Convention on the Prevention of Marine Pollution by Dumping of Wastes and Other Matter 1972), however abandonment must be for the "sole purpose of deliberate disposal" (Article 1.4.1.4) and does not include "abandonment in the sea of matter (e.g. cables, pipelines and marine research devices) placed for a purpose other than the mere disposal thereof" (Article 1.4.2.3).

16 Article 60(3), UNCLOS.

17 International Maritime Organization (IMO) (2008) *Guidelines and Standards for the Removal of Offshore Installations and Structures on the Continental Shelf and in the EEZ*. IMO Resolution A.672(16). London: IMO.

18 Article 60, UNCLOS.

19 The Guidelines state that consideration of environmental factors should be based upon scientific evidence taking into account the effect on water quality, geological and hydrographic characteristics, the presence of endangered or threatened species, existing habitat types, local fishery resources, and the potential for pollution or contamination of the site (section 2, paras.2.1.1–2.1.6).

20 The Guidelines outline the factors that should be considered in determining each of the preceding matters (section 2.3).

21 Section 3, para. 3.4.1. E.g.: OE devices may alleviate fishing pressure and potentially allow fish to breed and grow (Witt, M. J., et al. (2012) Assessing Wave Energy Effects on Biodiversity: The Wave Hub Experience. *Philosophical Transactions Series A. Math Physical and Engineering Sciences*, 370(1959), p. 502); introduce new hard substrate that may have artificial reef effects (Linley, E. A. S., et al. (2007) *Review of the Reef Effects of Offshore Wind Farm Structures and Their Potential for Enhancement and Mitigation*. London: Department for Business, Enterprise and Regulatory Reform (BERR), p. 132); or act as fish aggregating devices (Wilhelmsson, D., Malm, T., and Öhman, M. C. (2006) The Influence of Offshore Wind Power on Demersal Fish. *ICES Journal of Marine Science*, 63(5), p. 775).

22 These include the International Convention for the Safety of Life at Sea (SOLAS); Convention on the International Regulations for Preventing Collisions at Sea (COLREGS); the International Load Line Convention; the International Convention on Standards of Training, Certification and Watchkeeping for Seafarers (STCW); and the International Convention for the Prevention of Pollution from Ships (MARPOL).

23 GL Garrad Hassan Canada (2012) *International Overview of Marine Renewable Energy Regulatory Frameworks*. Ottawa: GL Garrad Hassan.

24 The International Association of Lighthouse Authorities (IALA) previously issued recommendations on the marking of a single offshore wind turbine (Recommendation O-114 in 1998), offshore wind farms (Recommendation O-117 in 2004) and wave and tidal devices (Recommendation O-131 in 2005). These have now been consolidated by IALA Recommendation O-139 on the Marking of Man-Made Offshore Structures (2008).

25 Articles 20 and 21. The IMO's General Provisions on Ships' Routeing (IMO Resolution A.572 (14), adopted on 20 November 1985) further expands on these provisions.

26 An undersea connection point for device testing.

27 Wave Hub. Available at: www.wavehub.co.uk/
28 The amendments were adopted by IMO's Sub-committee on Safety of Navigation at its 54th session, 30 June–4 July 2008 and approved by IMO's Maritime Safety Committee at its 85th session, 26 November–5 December 2008. The amendments entered into force on 1 July 2009. IMO (2008) Routeing of Ships, Ship Reporting and Related Matters. Amendments to the Traffic Separation Scheme 'Off Land's End, Between Longships and Seven Stones'. Submitted by the UK. Sub-Committee on Safety of Navigation, 54th session. IMO Doc. NAV 54/3/5, 28 March 2008. London: IMO.
29 Resolution A.671(16).
30 Wavenet (2003) *Results from the Work of the European Thematic Network on Wave Energy*. European Community ERK5-CT-1999-20001, 2000–2003.
31 Neumann, F. (2009) *Non-technological Barriers to Wave Energy Implementation*. Waveplam. Available at: http://www.youblisher.com/p/1276753-Non-technological-barriers-to-wave-energy-implementtion-Waveplam/
32 O'Hagan, A. M. (2012) n. 1; Kolliastas, C., et al. (2012) *Offshore Renewable Energy: Accelerating the Development of Offshore Wind, Tidal and Wave Technologies*. Abingdon: Earthscan; Leary, D., and Esteban, M. (2009) Climate Change and Renewable Energy from the Ocean and Tides: Calming the Sea of Regulatory Uncertainty. *International Journal of Marine and Coastal Law*, 24(4), pp. 617–651; Appiott, J., Dhanju, A., and Cicin-Sain, B. (2014) Encouraging Renewable Energy in the Offshore Environment. *Ocean & Coastal Management*, 90, pp. 58–64; Muñoz Arjona, E., et al. (2012) *Navigating the Wave Energy Consenting Procedure: Sharing Knowledge and Implementation of Regulatory Measures*. SOWFIA Project Third Workshop Report. Dublin: SOWFIA.
33 However, several OE test centres are "pre-consented", such that developers do not usually have to undertake the full consenting process.
34 Neumann, F. (2009) n. 31.
35 Ibid.
36 Subject to certain conditions and the right of innocent passage. At present, all OE projects are located close to shore, within the territorial seas of states (12 nm). OE projects could theoretically be developed in the exclusive economic zone (EEZ – 200 nm) and even areas beyond national jurisdiction (ABNJ), though this is not currently economically feasible. This may be more likely for certain technologies such as OTEC and submarine geothermal energy. A number of offshore wind projects are located in the EEZ, and projects are moving further offshore (e.g. in Belgium, Denmark and Germany). This should be considered in the context of international ocean governance processes that may affect future OE developments in ABNJ, see Abad Castelos, M. (2014) Marine Renewable Energies: Opportunities, Law, and Management. *Ocean Development & International Law*, 45(2), pp. 221–237.
37 See e.g. Nova Scotia and the UK, discussed in the following chapter; see also Wright, G. (2014) Regulating Marine Renewable Energy Development: A Preliminary Assessment of UK Permitting Processes. *Underwater Technology: The International Journal of the Society for Underwater*, 32(1), pp. 1–12.
38 Article 13(1)(c) and (a).
39 Article 13(1)(f).
40 Renewable Energy Progress Report COM (2013) 175 final.
41 Ibid.
42 White, R. (2012) Climate Change and Paradoxical Harm. In: Farrell, S., Ahmed, T., and French, D. (eds) *Legal and Criminological Consequences of Climate Change*. London: Hart Publishing.
43 For further discussion, see e.g. Sarmento, A. (2007) *Large Scale Environmental Impact Assessment of Wave Energy Devices – A Guidance Document*. Wavetrain; Leeney, R. H., et al. (2014) Environmental Impact Assessments for Wave Energy

Developments – Learning from Existing Activities and Informing Future Research Priorities. *Ocean & Coastal Management*, 99, pp. 14–22.

44 See Annex IV (Environmental Effects of Marine Renewable Energy) work and Tethys database. Available at: http://tethys.pnnl.gov/

45 EIA Directive (2014/52/EU), which entered into force in May 2014. Member States had until 16 May 2017 to transpose its provisions into their domestic legislation.

46 Article 4(6).

47 Articles 5(3)(a) and (b), respectively. No definition of what constitutes a "competent expert" is included in the text of the Directive.

48 Bailey, I., West, J., and Whitehead, I. (2011) Out of Sight but Not out of Mind? Public Perceptions of Wave Energy and the Cornish Wave Hub. *Journal of Environmental Policy and Planning*, 13(2), pp. 139–158; Chozas, J. F., Stefanovich, M. A., and Sørensen, H. C. (2010). *Toward Best Practices for Public Acceptability in Wave Energy: Whom, When and How to Address*. Third International Conference on Ocean Energy, Bilbao, Spain.

49 In the sense that, while OE resources are renewable, the number of optimal locations for OE deployment is limited.

50 Wright, G., et al. (2016) Establishing a Legal Research Agenda for Ocean Energy. *Marine Policy*, 63, pp. 126–134.

51 See e.g. HM Government (2010) *Marine Energy Action Plan 2010: Executive Summary & Recommendations*. London: DECC, p. 52. Available at: http://regensw.s3.amazonaws.com/1275819743_963.pdf; Freds Marine Energy Group (2009) *Marine Energy Road Map*. Edinburgh: Scottish Government, p. 61. Available at: www.gov.scot/Resource/Doc/281865/0085187.pdf; Jeffrey, H., and Sedgwick, J. (2011) ORECCA *European Offshore Renewable Energy Roadmap*. Edinburgh: University of Edinburgh on behalf of the ORECCA project, p. 101. Available at: www.orecca.eu/c/document_library/get_file?uuid=1e696618-9425-4265-aaff-b15d72100862&groupId=10129; Soerensen, H. C., and Rousseau, N. (2009) *Waveplam: Best Practice*. Bilbao: EVE, p. 47. Available at: www.youblisher.com/p/1276756-Best-practice-Waveplam/; Muñoz Arjona, E., et al. (2012) n. 32.

52 See Krueger, R. and Yarema, G. (1981) New Institutions for New Technology: The Case of Ocean Thermal Energy Conversion. *Southern California Law Review*, 54, p. 767.

53 Particularly in Denmark, where OSS is generally considered to have been a key driver of strong wind energy development. See Soerensen, H. C., and Naef, S. (2008) *Report on Technical Specification of Reference Technologies (Wave and Tidal Power Plant)*. NEEDS Integrated Project Report, p. 59. Available at: www.needs-project.org/2009/Deliverables/RS1a%20D16.1%20Final%20report%20on%20Wave%20and%20Tidal.pdf; Neumann, F. (2009) n. 31.

54 Muñoz Arjona, E., et al. (2012) n. 32.

55 Enrique, M. A., et al. (2012) *Navigating the Wave Energy Consenting Procedure: Sharing Knowledge and Implementation of Regulatory Measures*. Dublin: SOWFIA.

56 Scarff, G., Fitzsimmons, C., and Gray, T. (2015) The New Mode of Marine Planning in the UK: Aspirations and Challenges. *Marine Policy*, 51, pp. 96–102.

57 Simas, T., et al. (2015) n. 1.

58 Enrique, M. A., et al. (2012) n. 55.

59 See n. 52. The authors draw on earlier discussion of the concept by Humphreys in 1973. Humphreys, D. L. (1973) NEPA and Multi-Agency Actions – Is the 'Lead Agency' Concept Valid? *Natural Resources Lawyer*, 6(2), pp. 257–264.

60 E.g. the US Federal Energy Regulatory Commission offers an "integrated licensing process" that resembles the lead agency model.

61 Specifically, the Crown Estate Act 1961 states that TCE's duty in relation to the seabed is to "maintain and enhance its value and the return obtained from it, but

with due regard to the requirements of good management". Crown Estate Act 1961, section 1(3).

62 Such as the European Marine Energy Centre, Orkney and the WaveHub, Cornwall.

63 One in the Pentland Firth and Orkney waters strategic area (north of Scotland) and the other in the Rathlin Island and Torr Head strategic area (Northern Ireland).

64 In addition to the leasing process described here, TCE has initiated 40 technical studies in order to de-risk project development, published a study on MRE resources and conducted an industry engagement exercise on the future of the leasing process.

65 House of Commons Treasury Committee (2009) *The Management of the Crown Estate*.

66 Ibid.

67 Ibid.

68 Ibid.

69 Ibid.

70 Ibid.

71 In force since 6 April 2011.

72 The Secretary of State is the licensing authority for oil and gas–related activities and administers marine licences through the Department of Energy and Climate Change.

73 Section 66, Marine and Coastal Access Act 2009. In some cases, a marine licence is required for activities outside UK waters, e.g. where the activity takes place from a British vessel or where the vessel was loaded in UK waters.

74 Nationally Significant Infrastructure Projects, processed by the Planning Inspectorate, which makes recommendations to the Secretary of State to decide whether to grant consent (Section 15, Planning Act 2008). The MMO is a key consultee and remains responsible for monitoring compliance and enforcement of licence conditions under a Deemed Marine Licence.

75 Also entered into force on 6 April 2011.

76 Under the Marine and Coastal Access Act 2009, they are also the licensing and enforcement authority for the Scottish offshore region from 12 to 200 nm (other than reserved matters).

77 The Conservation (Natural Habitats) Regulations 1994. These were amended only for Scotland, while the Conservation of Habitats and Species Regulations 2010 applies across the remainder of the UK.

78 Marine Scotland aims to include this in their portfolio, allowing for consideration of onshore works.

79 Enrique, M. A., et al. (2012) n. 55.

80 Ibid.

81 Ibid.

82 Dominguez Quiroga, J. A., et al. (2013) *Report on Analysis of Existing Available Data of Wave Energy Experiences*. Seville: Abengoa Seapower on behalf of the SOWFIA project, p. 41.

83 Enrique, M. A., et al. (2012) n. 55.

84 Ibid.

Chapter 8

Consenting ocean energy projects

An overview of procedures in selected jurisdictions

Anne Marie O'Hagan and Glen Wright

1 Introduction

In this chapter, we provide an overview of consenting procedures in some selected jurisdictions: Australia, Canada, Denmark, France, Ireland, New Zealand, Portugal, Spain, the United Kingdom, and the United States of America. It is important to note that this chapter does not seek to present a detailed technical discussion of all aspects of the consenting processes in these jurisdictions. Rather, the brief outlines that follow are intended to provide an overview of the breadth and types of consenting process, as well as highlight some of the issues identified in the previous chapter.

2 Australia

Australia has a federal system of government, meaning that both Commonwealth and state environmental impact assessment (EIA) and planning legislation apply to ocean energy (OE) projects. There is currently no legislation specifically targeting OE.[1] All projects are governed by overarching legislation at the national level, in particular the Environment Protection and Biodiversity Conservation Act 1999, as well as legislation on the electricity market, renewable energy, and maritime safety. Projects within 3 nm (nautical miles) of the coast also come under the purview of state legislation. For example:

- In Victoria, OE projects require planning approval, environmental assessment and approval, works approval, infrastructure approval, heritage and Aboriginal impact approvals and electricity and gas approvals.[2] This multitude of approvals, coupled with the novelty of the technology and the lack of available scientific data, has resulted in developers having to "forge a process" for approval of their projects.[3] The Government of Victoria has considered the reform of their planning framework, both for renewable energy projects generally and for OE projects in particular,[4] though momentum towards reform appears to have stalled in recent years.
- In Western Australia, Carnegie's Garden Island wave energy project was located within an active defence base on Commonwealth-owned land/sea

and was therefore subject to federal legislation. In addition, approvals were also required under Western Australia state law, including the Environmental Protection Act 1986, the Navigable Waters Regulations 1958, and the Wildlife Conservation Act 1950.[5] The terrestrial elements of the project were managed by the Department of Defence that issued an Environmental Clearance Certificate for the project in November 2012. The purpose of this project was to demonstrate survivability of the CETO device over the period of four seasons. The demonstration achieved over 14,000 in-ocean hours over a 12-month period, and the devices have now been retrieved, with a subsequent project planned for 2017.[6]

3 Canada

Canada has a federal system of government. At the federal level, required approvals for an OE project include those relating to land use, project activities, electricity transmission, occupational health and safety, operational safety, environmental protection, and navigation.[7]

In 2011, the federal Government established the Marine Renewable Energy Enabling Measures (MREEM) programme, tasked with, inter alia, developing a federal policy framework for administering OE activities and presenting its recommendations to Cabinet. Natural Resources Canada is conducting extensive research and analysis on relevant federal legislation and regulations and on other OE regulatory regimes in other jurisdictions under this programme.[8] At the time of the last evaluation (2015), the MREEM programme had not yet delivered its recommendations to Cabinet and was preparing for consultations with stakeholders in other jurisdictions as well as within industry.[9]

3.1 Nova Scotia

In Nova Scotia, where there is considerable potential for development of tidal energy projects, the legislature has passed the Marine Renewable Energy Act (2015),[10] which aims to:

> provide for the responsible, efficient and effective development of marine renewable-energy resources through
>
> (a) a regulatory system that
>
>> (i) is staged, collaborative, consultative and adaptive, and
>> (ii) integrates technical, environmental and socio-economic factors; and
>
> (b) programmes and initiatives that promote the sustainable growth and management of the marine renewable-energy sector in the Province.

The Act tasks the Minister with promoting the sustainable development of OE resources, including through establishing policies and permitting processes, and identifies two priority areas (the Bay of Fundy and Cape Breton Island's Bras d'Or Lakes). The Act allows the provincial authorities to identify and designate certain parts of the two priority areas as "Marine Renewable-Electricity Areas" (MREA).[11] The Minister can issue a call for applications for projects within a MREA, and a proponent may apply for a license only under such a call. The Act provides that any license issued must contain terms focused on environmental protection, such as the requirement for a license holder to comply with an approved environmental monitoring plan and rehabilitation of license areas upon decommissioning.

The regulatory regime caters for both the private industry and research organizations working in the marine renewable energy sector and allows for issuance of permits to entities interested in testing generators and other related installations.

4 Denmark

While the conditions for offshore wind farms are set out in Denmark's Promotion of Renewable Energy Act (2008), there is no specific process or requirements for OE projects. The Act does, however, provide that the Danish State has the right to exploit energy from water and wind within territorial waters and the exclusive economic zone (EEZ).[12] Two processes can apply to offshore wind farm projects, namely a tender procedure operated by the Danish Energy Agency (DEA) or an open-door procedure. The DEA operates as an OSS for consenting of offshore projects.

In Denmark, three licenses are required to develop an OE project:[13]

1   License for preliminary investigations
2   License for an offshore site (given where preliminary investigations prove the project is compatible with the relevant marine area)
3   License to exploit and approval for electricity production (granted providing conditions in the license to establish the project site are met)

Various authorities are involved in consenting projects:

- The Danish Coastal Authority[14] works together with local government to manage coastal use and seabed in territorial waters.
- The Danish Maritime Authority[15] is responsible for maritime safety and the use of the sea.
- The Danish Environmental Agency[16] is responsible for environmental impact assessment (EIA).

The Danish Energy Agency coordinates these authorities as part of its OSS approach. This involves organising a meeting for all the other authorities with a statutory remit to discuss the project application. Based on the result of the meeting, the DEA will decide whether the proposal can go ahead: if this is the case, the DEA will then inform the applicant and issue a license for preliminary investigation, including an EIA.

Overall, the consenting process can take from a few months (e.g. for a single or test device at an unproblematic site) to several years for more complex projects.[17]

5 France

There is no specific consenting process for OE technologies in France, in spite of increasingly strong policy support for its development. An OE project generally requires a number of different consents:

1   An application to the relevant Maritime Prefecture[18] for the right to use the French public maritime domain (baseline to 12 nm)
2   An authorisation under the Water Act 2006[19] (Conditions attached to the authorisation under the Water Act can relate to cetaceans, EMF and noise monitoring.)[20]
3   An authorisation under the French Energy Code 2011 (Where a company has been selected following a call for tenders, authorisation is automatically granted.)[21]

The type of consent applied for will dictate the necessity of an EIA. Consents granted under the Water Act are subject to an impact study.[22] Consent applications pertaining to the Public Maritime Domain must be accompanied by an EIA.[23] The EIA required, along with the application for Public Domain consent, is administered by the Environmental Agency on behalf of the Préfet de Département.[24] The Environmental Agency makes recommendations on the submitted EIA, but the final decision is made by the Préfet de Département.

With respect to the terrestrial aspects of an MRE project, if they are located within the Public Maritime Domain, they are exempt from planning permission for all related terrestrial electrical works, including works related to grid connection.[25] In other words, planning permission is deemed to be granted under the maritime domain consent. The time taken to obtain all the consents is estimated to take 14 months from the submission of an official application.[26]

There is currently one operational wave energy test site in Le Croisic, Brittany (SEM-REV).[27] For its development, three licenses were acquired: a power generation permit (Ministry of Energy), a temporary concession for the occupation of a restricted zone, and an approval under the French Water Act.[28] Auxiliary consents were also obtained, including permits for coastal construction and terrestrial infrastructure. Power generation and grid connection aspects of the project proved challenging as this involved a separate administrative process administered by the French distribution grid operator.[29]

A central part of the consenting process in France is the creation of a regulatory pathway that facilitates the involvement of strategic stakeholders in the evaluation of a project proposal and subsequent granting of the required consents. All development projects over €300 million require a public debate.[30] Under the Public Maritime Domain Decree, this process operates similar to a public inquiry. As the project progresses, communication becomes more frequent, and dedicated meetings with stakeholders are organised on a regular basis.

6 Ireland

The consenting process in Ireland involves a number of different consents and authorities. The system is currently being reformed, and new legislation amending the process is expected to be proposed shortly. Currently, any development occurring on the foreshore[31] requires consent in the form of a Foreshore Licence and/or Lease.

- A Foreshore Licence, granted for specific activities and for time limited periods, such as site investigation works and EIA scoping. The developer will meet with departmental officials to scope the content of the EIA either prior to applying for a Foreshore Licence or once they have obtained the license, as it is usual for the same or similar information to be required. The preliminary stages of the EIA process allows the developer to consult with the adjoining planning authority, for example, as well as other interested parties and relevant regulators with a view to establishing what should be contained in the environmental impact statement.
- A Foreshore Lease to undertake further development activities,[32] conveying the developer with a right to exclusive use. A lease is generally granted for 35 years and is subject to regular review by the responsible authority. In order to obtain a Foreshore Lease for the construction phase, a developer must be able to show that other relevant licenses, permissions and approvals have been obtained from additional regulators.

The Department of Housing, Planning, Community and Local Government (DHPCLG) is currently responsible for foreshore consenting. No timelines are associated with decisions on Foreshore Licence and Lease applications. At all stages, the Minister reserves the right to reject any application for a Foreshore Licence or Lease, to modify the area sought under license or to allow others to simultaneously investigate the suitability of the license area.[33] The foreshore consenting process has long been recognised as in need of reform. In 2013, the Department published the Maritime Area and Foreshore (Amendment) Bill that sought to align the foreshore consent system with the land-based planning system, to provide for a single EIA for projects, and to provide a coherent mechanism to facilitate and manage development in the EEZ and on the continental shelf.[34]

An EIA is required if likely significant environmental impacts are anticipated. OE projects fall under Annex II of the EU EIA Directive, which provides that the need for an EIA is decided on a case-by-case basis. An Appropriate Assessment, derived from the EU Habitats Directive, may be required if a project is expected to have significant effects on the conservation objectives of a designated Natura 2000 site.[35]

OE devices qualify as electricity generating stations under the Electricity Regulation Act 1999 and as such are subject to a complex licensing and authorisation procedure based on installed capacity.[36] Grid connection is subject to a separate administrative process operated by the transmission system operator, EirGrid. Planning permission from the adjoining local authority (county council) under the Planning and Development Acts 2000–2010 is also required for projects that have a land-based element such as a substation or the like.

Ireland has an operational quarter-scale test site in Galway Bay, on the west coast of Ireland. This site is located 1.5 km offshore in water depths ranging from 20 to 23 m and has already hosted a number of devices since its launch, for example WaveBob and the OE Buoy. A full-scale test site, the Atlantic Marine Energy Test Site (AMETS) is also under development on the west coast. It is being developed by the country's energy agency, the Sustainable Energy Authority of Ireland (SEAI), to enable testing of full-scale wave energy converters in an open ocean environment. The site will be grid connected and have test berths at various water depths to suit different types of devices.

7 New Zealand

In New Zealand, the Crown does not own the marine environment or the seabed (called the "common marine and coastal area"), nor is it possible for the Crown or any other party to do so.[37] This means that the proponent of an OE project cannot own or lease from the Government the seabed or marine space where their project is located. An OE proponent can hold only a resource consent, as will be described. While there is no specific process for OE projects, the Resource Management Act 1991 (RMA) provides the core legal framework for environmental management. The RMA provides for a regionalised system, whereby national standards, regulations and policies are implemented and supplemented by regional policy statements and plans, made by regional councils.[38]

A proponent of an OE project within the marine and coastal area[39] must apply for a resource consent from the regional council.[40] A coastal permit, a type of resource consent, permits the holder to conduct activities in the coastal/marine area that would otherwise contravene the RMA. As some activities relevant to the development of OE projects are restricted by the RMA,[41] a coastal permit is required to develop an OE project. A coastal permit is also required to utilise the water itself.[42] The maximum duration of a coastal permit is 35 years,[43] though in practice consents are generally much shorter, with long-term consents generally being subject to onerous conditions.[44] The relevant regional council decides whether to charge a proponent for their occupation of the marine area, having regard to the public and private benefits conferred by the occupation. Charges may also be levied for rental of the seabed on which the cables lie.[45]

A number of prototype and test deployments have taken place in New Zealand, though the consenting process was first tested for large-scale projects with the Crest Energy project. Crest Energy applied to Northland Regional Council for consents under the RMA to establish an array of 200 tidal turbines on the seabed of the Kaipara Harbour, connected to a land-based electricity substation via two undersea cables.[46] The project proved controversial and a case resulted in the Environmental Court. While the planning provisions themselves were not the subject of significant debate,[47] the Court heard evidence and discussion on a number of contentious issues, including sustainable management; navigation; coastal planning processes; Maori cultural issues; marine life, fish and fisheries; and monitoring and adaptive management.[48] The two primary issues related to environmental concerns and interaction with an existing claim for customary ownership of the seabed.[49] The Court held two hearings, requested further expert advice to provide clarity, and required mediation between the parties.[50] Crest Energy were ultimately awarded a consent for the maximum period of 35 years, with the turbines to be installed in phases according to an environmental management plan taking an adaptive management approach. Overall, it took around five years to obtain consent for the project.

8 Portugal

Whilst there is no single piece of legislation to consent OE developments in Portugal, all applicable legislation has been modified to better suit the needs of the sector.[51]

The main consents required are as follows:

• A license or a concession to use the water resource from the Portuguese Environmental Agency (APA).[52] The process for both types of consent begins the same way: the proponent must submit a pre-application form describing the project, its location and the site characteristics.[53] Larger projects requiring a concession will additionally be subject to a competitive public examination.[54]
• The environmental license or EIA is processed by the local Coordination Committee for Regional Development, which is involved in the screening and scoping stages of the EIA and in designating the competent authority to assess the application.[55] An Environmental Impact Statement (EIS) is submitted to the competent authority, which then makes a recommendation to the Minister, who can conclude that the assessment is favourable, conditionally favourable or unfavourable.
• A license for a power production installation, administered by the Energy and Geology Directorate-General. Pre-existing legislation has been amended so

as to better accommodate renewable energy production.[56] Grid connection is requested separately from the Portuguese Electricity Utility.

• Terrestrial infrastructure is subject to planning legislation, with permission required from the adjoining municipal council.

Importantly for OE developments, the Portuguese Government has been developing an MSP system since December 2008 and as a result has introduced an electronic one-stop shop whereby information, applications and payments are processed in a centralised system.

9 Spain

In Spain, the consenting process for OE projects is based on three main legal instruments, which have been amended to streamline consenting.[57] Spain has implemented a parallel processing approach so as to allow applications to be processed simultaneously, though the following consents are still interdependent:[58]

• The Coast Law of 1988[59] governs occupation of marine space and safety of navigation. The General Council on Coast and Ocean Sustainability[60] has jurisdiction over the territorial sea, and OE projects must comply with administrative requirements for granting of titles for territorial occupation.

• Royal Decree 1/2008[61] provides for EIA. The developer must first submit a preliminary document outlining the project and its anticipated environmental impacts; the competent authority then decides whether a full EIA is required. Following the EIA, the Environmental Authority can decide to grant an Environmental Authorisation and set out project-specific conditions. Amendments to this legislation intended to overcome delays in the consenting process now mean that all OE projects are subject to a simplified EIA process (the Law imposes a statutory time frame for the delivery of environmental approval – no more than 4 months or 6 months in special circumstances).[62]

• Royal Decree 1028/2007 covers consenting for offshore installations producing electricity in the territorial sea.

• Royal Decree 1955/2000 covers general energy consenting and stipulates that all the electricity generating installations require:

  • Administrative Authorisation (AA): A technical document relating to the project's installation plan;

  • Project Execution Approval (AEP): Relating to the commissioning of the project and enabling the project proponent to start construction;

  • Exploitation Authorisation (EA): Once the project is constructed, this allows the development to be 'switched on' and proceed to commercial production.

There is no pre-application consultation phase in the Spanish consenting process, so OE developers enter directly into the relatively complex licensing system that involves a number of authorities.[63] The overall time frame needed for an OE project to be consented is approximately two years, though this time frame is yet to be verified in practice.[64] For consenting of the *bimep* test site (Biscay Marine Energy Platform), granting of the maritime public consent took four years in total.[65]

10 United Kingdom

The UK has a devolved system of government with specific responsibilities retained by the UK Government and others devolved to the governments of Scotland, Wales and Northern Ireland.

The Crown Estate (TCE) acts as the manager of Crown-owned lands, that is approximately half the foreshore and almost the entire seabed in the UK's territorial seas.[66] TCE has the power to issue leases, depending on the position of the site and type of renewable energy project in question. TCE Seabed Lease is a commercial agreement that enables the developer to construct and operate the project for a set period of time. In return for the lease granting the developer seabed rights, the developer pays TCE an agreed rent. The lease will also include a number of site-specific controls and restrictions on the developer and oblige them to comply with other statutory consents. The leasing process is as follows:[67]

1   **An exclusivity agreement.** A reservation over an area of seabed or over seabed rights is granted that guarantees that TCE will not permit another developer (of the same technology) to use that area. During this time, a developer may conduct environmental and engineering studies so as to better define the project area. The exclusivity agreement also specifies how the developer can avail of the option to call for an Agreement for Lease, subject to meeting certain milestones.
2   **An Agreement for Lease (AfL).** Conditional on the developer having obtained the necessary statutory consents for the project, an AfL is then concluded that provides a developer with an enforceable option to call for a lease from TCE. According to TCE guidance, this restricts the ability of TCE to grant rights over the area to other users and confers on the developer:

    - Exclusivity over an area of seabed;
    - An enforceable option to require TCE to grant a lease;
    - Temporary rights to conduct ancillary activities relating to the project; and
    - Rights to call for the addition of an export cable route.

Following the conclusion of an AfL, the developer must obtain the necessary statutory consents from the relevant authorities (these requirements vary by jurisdiction; see later in the chapter). The developer first conducts all necessary technical and environmental work at the site/project level before continuing with the statutory consenting process under other legislation. During this period, the developer has no right to begin construction or operation of the project. Once the necessary approvals have been obtained, TCE will then award a seabed lease to the developer.

10.1 England and Wales

The UK's Marine and Coastal Access Act 2009 sought to modernise and reform marine planning and licensing arrangements and took significant steps to streamline consenting processes, including through the establishment of a one-stop shop, the Marine Management Organisation (MMO).

In England, the MMO acts as the decision maker on behalf of the Secretary of State for Business, Energy and Industrial Strategy (previously Secretary of State for Energy and Climate Change) for offshore developments generating up to 100 MW. For projects larger than 100 MW, the Infrastructure Planning Commission (IPC) makes the decisions on behalf of the Secretary of State. In Wales, Natural Resources Wales is responsible for granting marine licenses in inshore waters (up to 12 nm). Beyond 12 nm, the MMO grants such licenses.

A marine license covers the construction of works in the marine environment, the deposit of substances or articles and dredging.[68] To build and operate an energy generation installation a Section 36 consent is required under the Electricity Act 1989.[69] In England, developers can apply to the MMO for a safety zone declaration around offshore renewable energy generating stations.[70] The MMO process consists of three main stages. After registering for an online service account, the developer must submit:

1 An initial enquiry;
2 A request for pre-application advice (e.g. screening, scoping, environmental statement (ES) review); and, finally,
3 The license application.

Depending on the location of an OE project, a developer may also require a European Protected Species license under the Conservation of Habitats and Species Regulations 2010, while local planning authorities or councils are responsible for terrestrial infrastructure.

10.2 Scotland

The Scottish Government has also attempted to refine its consenting processes to better reflect the needs of the OE industry. In Scotland, the Marine and Coastal Access Act 2009 was supplemented by the Marine (Scotland) Act 2010, which provides for a single marine license and a streamlined application process.[71]

The Scottish Act also created a pre-application consultation process where prospective applicants must submit a pre-application notice and report before submitting their application.

The Marine Scotland Licensing Operations Team (MS-LOT) is responsible for all marine licensing functions. Marine Scotland operates as a one-stop shop, so there is a single point of contact to administer project development applications. Marine Scotland has a policy target of providing a decision on all applications within nine months (provided the requisite environmental information is available and no local public inquiry is needed). In practice, the consenting process has generally taken longer.[72]

Marine Scotland has published a licensing manual to assist developers in navigating the process,[73] which in addition sets out all additional requirements (e.g. those deriving from EU legislation). In practice, when a developer has a pre-application consultation with MS-LOT, they will advise the developer on the suitability of their site, whether an AA may be needed, information on the type of environmental data that is required, the stakeholders to be consulted and the expected costs and time frames associated with the process.

After the pre-application consultation, the developer will commence the EIA process and stakeholder consultation. This information, coupled with Navigational Risk Assessment, Habitats Regulations Appraisal, and any other information requested, is then submitted to MS-LOT with a completed application form and payment. MS-LOT administers the applications and consultations necessary for the various consents and compiles the feedback and input obtained from statutory consultees and other stakeholders. At this point, if information is missing or additional studies are required, the developer will be requested to provide that. If the application is unsuccessful, MS-LOT will give the developer the reasons as to why this occurred and also advise the developer on how to proceed. When an application is successful, the various consents will be granted. It is normal for the consents to include standard terms and conditions. Such conditions usually apply to that specific development site and any ongoing or post-consent environmental monitoring requirements.

10.3 Northern Ireland

The consenting process for OE in Northern Ireland involves a lease from TCE, as in the other UK jurisdictions, and the following consents:

- A Marine License from the Department of Agriculture, Environment & Rural Affairs (DAERA) Marine and Fisheries Division under the provisions of the UK Marine and Coastal Access Act and Marine Act (Northern Ireland) 2013[74]
- Electricity generation consents from the Department for the Economy and the utility regulator[75]

- A license from the Utility Regulator under the Electricity (Applications for Licences and Extensions of Licences) (No. 2) Regulations (Northern Ireland) 2007
- The Marine Works (Environmental Impact Assessment) Regulations 2011, which consolidates previous requirements and brings together deposits, navigational activity, harbour works and marine minerals dredging in one legal framework

Depending on the complexity of the proposed project, the application is processed within four months from receipt of a completed application, provided it includes all the relevant information, supplementary material and correct fee. Licensing fees in 2016 ranged from £4,420 (Band R1. projects up to 0.99 MW) to £33,709 (Band R4, projects of 100 MW and above).[76] In line with the Northern Ireland Executive's commitment to marine renewable energy, the Marine Strategy and Licensing team in DAERA facilitate engagement between all relevant parties and statutory authorities in relation to specific requirements of the marine licensing process and can also assist developers in preparing their environmental impact assessments.

11 United States of America

In the United States, the Federal Energy Regulatory Commission (FERC) exercises regulatory jurisdiction over OE projects out to 3 nm.[77] Beyond 3 nm, that is in what is termed the 'outer continental shelf' (OCS),[78] the Bureau of Ocean Energy Management (BOEM) administers the leasing process. The construction and operation of an OE project on the OCS also requires a license from FERC.[79] Recognising that complex jurisdictional issues impacted the development of OE projects, both authorities published comprehensive *Guidelines on Regulation of Marine and Hydrokinetic Energy Projects on the* OCS in 2012.[80]

Three types of leases can be granted by BOEM for OE projects: a commercial lease; a research lease, issued only to federal agencies or states for research activities; and a limited lease, which applies to projects of limited scope.[81] BOEM customarily issue commercial leases for a period of 25 years, though this can be extended or amended to align with the associated FERC license. FERC licenses can be for up to 50 years, with supplementary powers to extend such a license for between 30 and 50 years.[82]

Leasing takes place through competitive rounds instigated by BOEM. If there is no suitable licensing round, a developer can submit an unsolicited application to BOEM stipulating their interest in obtaining a lease for that specific OCS location. If there are no other applications for that site,[83] the developer will then be requested to provide a site assessment plan (SAP), detailing environmental surveys and resource assessment studies to support the planned project.[84]

Whilst the granting of a lease and approval of an SAP enable a developer to conduct certain activities identified therein, the lease does not cover a right to generate power. The latter right is only conveyed once a FERC license has been obtained. According to the guidelines, where a BOEM lease and FERC license are required, the two processes can be combined in the interests of time and expediency. This is dependent on the type of license being applied for and whether it is in response to a competitive leasing round or an unsolicited application. The FERC licensing process can take one of three forms: an integrated licensing process (ILP), a traditional licensing process (TLP) and an alternative licensing process (ALP).[85]

The ILP is the most commonly applied process and involves a pre-application stage, during which any necessary studies are conducted and a license application is prepared, and a post-application stage, when the application is reviewed, an environmental document is compiled and a decision on licensing is made. During both stages, FERC will seek the input of stakeholders. In terms of applicable time frames, the *Guidelines* specify that for a BOEM lease in an area of competitive interest, the process could take up to 2.5 years, including stakeholder consultation and environmental work. Where there is no competitive interest, the time frame is less, perhaps one to two years depending on the complexity of the project being proposed. Receipt of a FERC license takes approximately one year, but this is largely dependent on the extensiveness of the application submitted. Pilot project licenses from FERC take six months from the date of application submission.

Occupation of the sea space under a commercial or limited lease requires an initial, one-off payment to BOEM to obtain the lease as well as ongoing, annual payments once the term of the lease begins. The amount payable to BOEM varies according to whether the developer is responding to a competitive leasing auction or a non-competitive leasing process. BOEM does not require payment for research leases. FERC require payment for the administrative charges associated with licenses for projects over 1.5 MW. They also need payment for land charges that vary according to whether the land is federal or tribal.

In addition to the core BOEM and FERC processes, other authorisations may be required, including approval for a structure to be deployed in navigable waters[86] and permits for dredging and other preparatory works.[87] The National Environmental Policy Act (NEPA) provides the framework for identifying and assessing the environmental effects, which applies to OE projects.

12 Conclusion

It is clear from the selected examples that there is widespread geographical variation in how OE projects are consented. This is an evolving field, and it is anticipated that this will continue to be the case until such time as OE becomes a fully commercial-scale sector. Consenting processes will also evolve in response to changing wider law and policy frameworks, such as through the implementation of integrated maritime governance and processes such as marine spatial planning.

It is clear from the preceding examples that most jurisdictions do not have specific OE consenting legislation and processes in place. This is not surprising given that OE could be viewed as merely another maritime activity, and hence it should be subject to the same level of regulation and control as already established marine industries. Where this becomes problematic is when you consider the current technology readiness levels of wave and tidal energy devices. At the time of writing, there are little or no commercial-scale wave energy deployments, and tidal energy deployments are just tipping towards commercial scale, with full-scale devices being deployed on a phased basis. It is arguably unfair and uneconomic to subject short-term deployments of single devices to the same rigours of a consenting process designed for multi-megawatt deployments of numerous devices for commercial generation. There is a need for a proportionate approach in both process and subsequent contractual terms and conditions.

Currently, consenting processes vary primarily due to the governance and institutional structures in place in a particular country. This can result in a division of responsibility and changes in procedure according to central and local government levels, between Commonwealth and state entities or according to land and sea elements of a project or a combination of both. This makes the process difficult to navigate, as well as expensive and lengthy. It may not be possible to overhaul the governance and institutional structures in place, but there is a definite need for dedicated coordination functions and responsibilities, as well as for a streamlining of consents. Some countries are at the vanguard of this; for example, Scotland has invested significant resources in making their procedures more efficient, effective and user-friendly. Spain, Portugal and Denmark have also attempted to better coordinate their multiple processes.

Although the implementation of a more coordinated, or one-stop shop, approach is being rolled out across numerous jurisdictions globally, it is often only instigated for the marine and environmental elements associated with OE developments, such as occupation of the sea space and offshore EIA work. Responsibilities relating to the electrical aspects of a development, grid connection, planning permission and associated terrestrial EIAs still appear to reside in different departments or authorities. As long as this situation continues, it will be difficult to expedite consenting.

Examples from various jurisdictions around the world indicate that the applicable legislation governing consenting is, in the majority of countries, capable of dealing with new technologies such as wave and tidal. Persistent problematic issues, however, relate to how competent authorities implement the legislation, how they work collaboratively and how they approach environmental considerations. The introduction and implementation of new management approaches, such as marine spatial planning, have the potential to influence and help streamline future procedures and processes, but as yet this seems to be having little impact. Whilst some countries do not have MSP in place, those countries that do have marine plans rarely reflect OE developments or the future needs of the sector. This could be ascribed to the current lack of deployments, but it may also be attributed to inconsistent messages from the sector.[88]

In conclusion, the key needs to advance the OE sector are better coordination and streamlining of processes and responsible authorities. Competent authorities and policymakers should also endeavour to ensure that consenting systems complement existing or forthcoming MSP. It may not yet be clear what form MSP will take, but the designation of specific zones for OE, for example, could be preconsented or require less environmental characterisation prior to devices being deployed. This type of approach may allow developers to progress from test centres, for example, to a more advanced stage within the same jurisdiction. Predefined zones could also form the basis of lease areas (precursor to a leasing round) for consenting authorities and also encourage developers to specific locations.

Notes

1 Ocean Energy Systems (OES) (2014) *Annual Report: Implementing Agreement on Ocean Energy Systems (IEA-OES)*. Lisbon: OES Executive Committee, p. 148. Available at: https://report2014.ocean-energy-systems.org/
2 Department of Primary Industries, briefing with the Environment and Natural Resources Committee – Melbourne, 22 June 2009. Quoted in Environment and Natural Resources Committee (2010) *Final Report for the Inquiry into the Approvals Process for Renewable Energy Projects in Victoria*. Melbourne: Parliament of Victoria. Available at: www.parliament.vic.gov.au/enrc/inquiries/article/870
3 Ibid.
4 Ibid.
5 Davies Ward, E. H. R., Wykes, B., Bailey, M., Sawyer, T. C., Development and Consenting of Carnegie Wave Energy's Perth Wave Energy Project, Experiences from Down Under (EIMR Conference, Stornoway, 30 April–1 May 2014). Available at: https://tethys.pnnl.gov/sites/default/files/publications/EIMR-2014-Abstract-Ward.pdf
6 Carnegie Wave Energy (2016) *Perth Wave Energy Project Update, March 2016*. Available at: https://s3-ap-southeast-2.amazonaws.com/website-sydney-1/media/2017/06/30111822/160202_Renew-Economy_Carnegie-completes-final-milestone-for-CETO-5-Perth-wave-energy-project.pdf
7 WavEC (2015) *Consenting Processes for Ocean Energy on OES Member Countries*. Lisbon: WavEC for the OES under Annex I – Review, Exchange and Dissemination of Information on Ocean Energy Systems. Available at: https://tethys.pnnl.gov/sites/default/files/publications/OES-AnnexI-Report-2015.pdf
8 Cox & Palmer (2016) *Nova Scotia: Marine Renewable Energy Act – A Synopsis*. Available at: www.coxandpalmerlaw.com/en/home/publications/nova-scotia-marine-renewable-energy-act-a-synopsis.aspx
9 Natural Resources Canada (2015) *Evaluation Report: Renewable Energy Deployment Sub-program (RED)*. Available at: www.nrcan.gc.ca/evaluation/reports/2015/17933
10 Nova Scotia Legislature (2015) *Marine Renewable-energy Act*. Available at: http://nslegislature.ca/legc/bills/62nd_2nd/3rd_read/b110.htm
11 The Act itself identifies four areas as MREAs for tidal projects and allows the Province to identify further priority areas and MREAs.
12 Article 3(1), Promotion of Renewable Energy Act, No. 1392 of 27 December 2008.
13 PNNL/US DOE/OES. 2015. *Tethys – Regulatory Frameworks*. Available at: https://tethys.pnnl.gov/regulatory-frameworks
14 An agency under the Ministry of the Environment.
15 An agency under the Ministry of Business and Growth.
16 An agency under the Danish Ministry of the Environment.
17 WavEC (2015) n. 7.

18 A civil servant who exercises authority over the sea in one particular region known as a *Préfecture maritime*.

19 République Française (2017) *Legifrance.gouv.fr – le service public de la diffusion du droit. Loi n° 2006–1772 du 30 décembre 2006 sur l'eau et les milieux aquatiques.* Available at: www.legifrance.gouv.fr/affichTexte.do?cidTexte=JORFTEXT000000649171

20 If the project involves investment of over €1.9 million, a public inquiry is mandatory. See Kablan, Y., and Michalak, S. (2013) *Politiques publiques de soutien aux projets d'énergies marines renouvelables (EMR) et cadres réglementaires régissant le secteur en France. Livrable 4.1.1 du WP4 du projet MERiFIC: Un rapport préparé dans le cadre du projet MERiFIC.* Truro: Cornwall Council, p. 78.

21 Article L311–6 of the 2011 code de l'énergie. See Kablan, Y., and Michalak, S., (2013) ibid.

22 Article R 214–6, Environment Code.

23 As regulated by article L 122–1 and R 122–2 and in line with the Environment Code.

24 Le Lièvre, C. and O'Hagan, A. M. (2016) *Legal and Institutional Review of National Consenting Processes Deliverable 2.2 RiCORE Project.* Cork: UCC, p. 54. Available at http://ricore-project.eu/wp-content/uploads/2016/02/RiCORE-D2.2-Legal-Institutional-Review-Final-1.pdf

25 Under the Law Grenelle II and Decree n° 2012–40, and the French Town Planning Code, articles L 421–5 and R 421–8–1.

26 This time frame refers only to the consenting process described here; however, an initial consultation process, during which EIA and other feasibility studies are conducted, starts one year before an application is submitted. In some cases, the EIA can take up to two years, depending on the scale and location of the project, though efforts have been made in recent years to streamline the EIA process.

27 Central Nantes SEM-REV–LHEEA. Available at: www.semrev.fr/

28 Simas, T., et al. (2015) Review of Consenting Processes for Ocean Energy in Selected European Union Member States. *International Journal of Marine Energy*, 9, pp. 41–59.

29 Ibid.

30 Article R 121–2, Environment Code.

31 The foreshore is legally defined as the area between mean high water and the 12-nm territorial sea limit. The original Foreshore Act was enacted in 1933, and, whilst it has been amended on numerous occasions to take account of other legislative changes, it remains the key instrument governing all marine and coastal development.

32 While it is not possible to obtain a Foreshore Lease unless the preliminary works have been conducted under a Foreshore Licence, successful completion of site investigation works does not automatically entitle a developer to a Foreshore Lease.

33 O'Hagan, A. M., and Lewis, A. W. (2011) The Existing Law and Policy Framework for Ocean Energy Development in Ireland. *Marine Policy*, 35(6), pp. 772–783.

34 Department of Housing, Planning, Community and Local Government (2016) *General Scheme of the Maritime Area and Foreshore (Amendment) Bill.* Available at: www.housing.gov.ie/planning/foreshore/general-scheme-maritime-area-and-foreshore-amendment-bill

35 For this to be the case, the proposed development does not necessarily have to be situated within the boundaries of the designated site but merely may arise when a proposal is adjacent to such a site.

36 O'Hagan, A. M., and Lewis, A. W. (2011) n. 33.

37 The Marine and Coastal Area (Takutai Moana) Act 2011 states "[n]either the Crown nor any other person owns, or is capable of owning, the common marine and coastal area" (s. 11(2)). That Act repealed the controversial Foreshore and Seabed Act 2004, thereby restoring customary title (s. 6(1), Marine and Coastal Area (Takutai Moana) Act 2011). Where a project is consented, the devices in the common marine and coastal area are treated as personal property.

38 Resource Management Act (RMA), s. 65(2) and Schedule 1. Regional policy statements "set the basic direction for environmental management in the region", while

regional plans "tend to concentrate on particular parts of the environment, like the coast, soil, a river or the air". Ministry for the Environment (2009) *An Everyday Guide to the RMA: Getting in on the Act*, Series 1.1, 2nd ed. Wellington: Ministry for the Environment. Available at: www.mfe.govt.nz/sites/default/files/media/RMA/RMA%20Booklet%201.1.pdf

39 Article 9(1) of the Marine and Coastal Area Act 2011 defines the marine and coastal area as the area from MHWS to the 12-nm limit.

40 Part 6A of the RMA.

41 For example, a person may not "erect, reconstruct, place, alter, extend, remove, or demolish any structure" that is "fixed in, on, under, or over any foreshore or seabed" (Section 12(1)(b) RMA) or disturb the foreshore or seabed in a way "likely to have an adverse effect" (Section 12(1)(c) RMA).

42 Section 14(1)(a) of the RMA states that a person may not "take or use (. . .) energy from any open coastal water" without consent.

43 Section 123(c) RMA.

44 Simpson Grierson (2005) *Marine Energy Proposals – Resource Management Act 1991*. Referenced in: Wright, G., and Leary, M. (2011) Marine Energy. *New Zealand Law Journal*, (August 2011), pp. 227–230.

45 Under the Resource Management (Transitional Fees Rents and Royalties) Regulations 1991 (s.8 and sch.2).

46 *Crest Energy Kaipara Limited & Others v. Northland Regional Council* [2009] A132 18 June 2009 (New Zealand Environment Court); para. 1.

47 Ibid. paras. 17–19.

48 Ibid. para. 34.

49 Ibid. para. 170. The Te Uri o Hau Settlement Trust argued that granting resource consents for the project would prejudice their prior claim lodged with the High Court.

50 Wright, G. (2011) A Tidal Power Project. *New Zealand Law Journal*, (September 2011), pp. 260–261.

51 Simas, T., et al. (2015) n 28.

52 Whether a lease or a licence is required depends on the scale of the device and the length of deployment: projects with a capacity of 25 MW or less require a licence, as do deployments of less than one year, whereas projects with a greater capacity or longer-term deployment require a concession.

53 The latter includes characteristics relating to navigation, fisheries, recreational areas, physical environment, meteorological conditions, emergency plans and any onshore elements associated with the project.

54 Simas, T., et al. (2015) n. 28.

55 An EIA is not required if a project is located outside the National Ecological Reserve, Natura 2000 sites or the national network of protected areas, but this must first be confirmed by the competent licensing authority. See Ministério da Agricultura, Mar, Ambiente e Ordenamento do Territóriio (2012) *Plano de Ordenamento do Espaço Marítimo (POEM): Volume Síntese Memória Geral da Proposta de POEM*. Available at: www.dgpm.mam.gov.pt/Documents/POEM_VolumeSintese.pdf

56 Decree Law 225/2007.

57 Simas, T., et al. (2015) n. 28.

58 Le Lièvre, C., and O'Hagan, A. M. (2016) n. 24.

59 Amended by Law 2/2013.

60 Part of the Ministry of Rural, Marine and Natural Environments.

61 Amended by Law 21/2013.

62 Ibid.

63 Le Lièvre, C., and O'Hagan, A. M. (2016) n. 24.

64 Ibid.

65 O'Hagan, A. M., et al. (2016) Wave Energy in Europe: Views on Experiences and Progress to Date. *International Journal of Marine Energy*, 14, pp. 180–197.
66 Under the Energy Act 2004, TCE hold the rights to the resources to develop renewable energy within the area defined as the renewable energy zone (REZ), which extends out to the continental shelf.
67 The Crown Estate (2014) *The Crown Estate Role in Offshore Renewable Energy Developments: Briefing (January 2014)*. London: The Crown Estate, p. 5.
68 Section 66, Marine and Coastal Access Act 2009.
69 Some electricity generating installation may be subject to regulation under the Electricity Works (EIA) (England and Wales) Regulations 2000 (as amended) which would necessitate an EIA to be submitted to the MMO with the main project application.
70 Section 95, Energy Act 2004.
71 This replaced the licences previously required under the Food and Environment Protection Act 1985 (FEPA) for deposits and under the Coast Protection Act 1949, relating to navigation. A licence to construct an electricity generating station is still required under section 36 of the Electricity Act 1989, but the application for this can now be considered at the same time as the application for the marine licence for an ocean energy project.
72 Wright, G. (2016) Regulating Wave and Tidal Energy : An Industry Perspective on the Scottish Marine Governance Framework. *Marine Policy*, 65, pp. 115–126.
73 Scottish Government (2012) *Marine Scotland Licensing and Offshore Wind Energy Development* (Report R.1957). Edinburgh: Scottish Government.
74 Part 4. This covers the marine elements of a project including the placement of anything on or removing material from the seabed.
75 Article 39, Electricity (NI) Order 1992 enables the construction and operation of generating stations.
76 See DAERA (2017) *Marine Licensing (Northern Ireland) Application Fees 2016*. Available at: www.daera-ni.gov.uk/publications/marine-licensing-northern-ireland-application-fees-2016
77 Under an amendment to the Federal Power Act authorising it to regulate and licence hydroelectric projects on navigable waters within 3 nm of the shore and any projects with an onshore grid connection. FERC does not have the powers to regulate OTEC projects, which fall under the remit of the National Oceanic and Atmospheric Administration (NOAA) according to the OTEC Act 1980.
78 Includes all submerged lands, subsoil and seabed between the seaward extent of the states' jurisdiction, usually 3 nm, to the limit of federal jurisdiction of 200 nm.
79 Under the Guidelines, a developer can develop a project under a BOEM lease but without a FERC licence if the technology is experimental; the proposed installation is temporary in preparation for a further licence or for educational purposes; and the power generated would not be transmitted to the grid. In certain other cases, FERC is authorised to grant licence waivers or exemptions.
80 Bureau of Ocean Energy Management and Federal Energy Regulatory Commission (BOEM/FERC) (2012) *BOEM/FERC Guidelines on Regulation of Marine and Hydrokinetic Energy Projects on the OCS*. Version 2, 19 July 2012. Available at: www.ferc.gov/industries//hydropower/gen-info/licensing/hydrokinetics/pdf/mms080309.pdf
81 Ibid. Generally, a project of no longer than five years and restricted capacity specified in the terms and conditions attached to the lease.
82 Though pilot projects licensed by FERC have a much shorter existence, usually five years so as to reflect the early stage of the technology, the scale of the project and its overall experimental nature.
83 For locations where there is a competitive interest in MHK project development in a specific area, BOEM will publish a Proposed Sale Notice followed by a Final Sale

Notice, once the requisite environmental review and consultations have been completed. Developers must comply with the Final Sale Notice. Once a developer has completed the necessary documentation and made payment, BOEM will issue a lease to the successful developer. The successful developer will have to submit a SAP within six months of issuance of the lease.

84 There is no requirement to submit a SAP if the project proposed does not involve the installation of bottom-founded facilities.
85 BOEM (2012) n. 81.
86 From the US Army Corps of Engineers under the Rivers and Harbors Act §10. If a device or any of its constituent parts might obstruct navigation, it must be marked by the appropriate navigational aid. This requires a Private Aid to Navigation (PATON) permit administered by the US Coast Guard under Title 33 of the Code of Federal Regulations.
87 Clean Water Act, §404.
88 O'Hagan, A. M. (2016) *Consenting Processes for Ocean Energy – A Report Prepared on Behalf of the IEA Technology Collaboration Programme for Ocean Energy Systems (OES)*. Lisbon: IEA-OES Secretariat, p. 40. Available at: www.ocean-energy-systems.org/library/oes-reports/consenting/document/consenting-processes-for-ocean-energy-barriers-and-recommendations-2016-/

Chapter 9

Ensuring the sustainable development of ocean energy technologies through environmental assessment laws and policies

Glen Wright, Edward Willsteed and Anne Marie O'Hagan

I Introduction

Environmental assessments (EA) are a range of tools commonly used in the regulation and consenting of projects and plans to estimate and evaluate the effects of a course of action on the environment. EA helps project proponents and decision makers consider the environmental consequences of proposed actions by identifying impacts and considering ways to minimise, mitigate or compensate for them.[1]

There are various forms of EA, the most established being the environmental impact assessment (EIA) conducted at project or site level. In many jurisdictions around the world, EIA has become the main tool utilised by regulatory authorities to account for the environmental consequences of individual projects relative to environmental protection goals. EIA is also one of the key interfaces between science and regulation as it can encourage both the utilisation of existing knowledge and the generation of new scientific data. Over time, EIA can increase knowledge about environmental interactions, thereby reducing uncertainty and ultimately resulting in better policy and decision making.

The emerging practice of strategic environmental assessment (SEA) is generally broader in focus and is conducted at a strategic level on a programme or plan. In the EU, there is also the appropriate assessment, a requirement under the Habitats Directive, which seeks to ascertain that a proposed project will not adversely affect the integrity of the designated site concerned.

This chapter provides an overview of the various forms of EA and their role in connecting science to the regulatory process, utilising two ocean energy (OE) case studies. The dichotomy between precaution and risk in project-level decision making is discussed. The chapter then discusses three approaches to better balance environmental protection with knowledge generation and innovation through the EIA process: adaptive management; survey, deploy and monitor; and the Rochdale Envelope. A brief outline of marine spatial planning (MSP) is discussed in the OE context.

The chapter concludes that application of EA processes must be adapted if they are to better serve marine management and balance competing goals. An adaptive EA process is needed to utilise existing data and factor it into decision-making processes, while also encouraging the generation of new information. It is argued that the incorporation of the more innovative approaches to management discussed can achieve this by providing for environmental protection in the absence of complete scientific certainty, while also allowing for the generation of new knowledge through considered, risk-based permitting and deployment of OE devices. Strategic and integrated marine governance initiatives can play a complementary role in filling knowledge gaps and reducing scientific uncertainty regarding the effects of new technologies in the marine environment.

2 Environmental impact assessment

The first formal EIA system was established in the United States in 1970 by legislation that was primarily a political response to the changing nature and scale of post–World War II development.[2] Public interest and concern regarding the environmental impacts of development was growing, and the adequacy of existing decision-making tools was criticised.[3] EIA is now a well-developed concept in environmental law, having been adopted in over 100 jurisdictions and in many bilateral and multilateral aid and funding agencies.[4] While the exact nature of the process can vary, EIA generally follows a series of similar stages:

1   **Screening** the proposed development to determine whether there are likely to be significant effects on the environment and whether a full EIA is needed.
2   **Scoping** the available environmental data and key issues at the proposed deployment site. The alternatives[5] to the project should be included in the scoping stage. This identifies additional studies that will be required in order to make a proper assessment.[6]
3   **Baseline studies** provide baseline data on the status of the receiving environment. At this stage, studies identified during the scoping phase will also be undertaken.
4   **Submission of environmental information to competent authority**, whereby the proponent compiles a statement or report of the anticipated effects and supporting documentation.
5   An **assessment of impacts** is then made to assess the significance of potential identified interactions. Options to mitigate these impacts are considered, and any residual effects are detailed.
6   **Decision:** On the basis of the foregoing information, consenting and regulatory bodies consider the documentation submitted and take this into account before making their decision on whether to permit the project and under what conditions.[7] The conditions are usually included with the license, lease or other consent that the development requires to progress.

EIA forms a key part of the regulatory process for OE in most, if not all jurisdictions. Existing EIA legislation is likely to be applied unaltered to OE proposals, though some countries have enacted amended EIA legislation to specifically take OE into account (e.g. Portugal). This is a common pattern with new technologies: they are developed in the context of existing legal frameworks but inevitably elicit new legal responses as commercialisation approaches.[8] Some specific potential responses to OE development are considered later in this chapter.

Despite widespread adoption and the considerable literature on the topic, EIA has only had limited success in advancing environmental protection. Indeed, the results "appear most favourable when compared with past neglect and failings, rather than when measured against sustainable development goals".[9] This limited success may stem from EIA's poor theoretical basis: while it is a quasi-scientific process, it emerged from a political rather than a scientific imperative. EIA practice therefore began before there was adequate scientific capacity to support it.[10] Though broader theoretical conceptualisations of EIA were later developed, building on evaluative research and appropriating theoretical frameworks from other disciplines,[11] overall, the theoretical underpinnings of EIA are an uneven and inconsistent mixture of science, planning, social, economic and biological theories, with the process as a whole often being less that than the sum of its parts.[12]

This lack of theoretical direction has given rise to a number of different conceptions of EIA, how it does and should operate, and the role of science in the process.[13] The approach to the role of science in EIA can be separated into two overlapping paradigms: EIA as applied science and EIA as civic science.[14] Under the former conceptualisation, EIA is a process by which objective scientific knowledge is brought to bear on practical decisions. By contrast, the latter encompasses diverse opinions, with a unifying belief that EIA is a tool for "influencing decisions through the application of a pragmatic, inclusive and deliberative form of science".[15] In the OE context, EIA has both relied on pure science to inform project approvals but also drawn on the participatory nature of the EIA as a civic science approach. Indeed, these are not strict divisions, and the flexibility and adaptability of EIA are part of its appeal as much as they are its weakness.

2.1 EIA and ocean energy projects

2.1.1 Current status of science/knowledge and regulation

OE technologies suffer from knowledge deficiency on three levels. Firstly, while other renewable energy technologies are well established,[16] there is limited practical experience with the deployment of OE technologies. Secondly, the marine environment is notoriously difficult to study, with research being limited by cost, inaccessibility and relative lack of development in comparison to the onshore environment. Impact assessments in the marine environment have been called "the most challenging of all".[17] Thirdly, the OE industry is developing at a time when a fledgling transition to more integrated marine management is underway, whereby legislation requires regulators and managers to take into account the cumulative effect of all maritime activities on ecosystems, as well as to consider future activities and environmental trends such as climate change.[18] Knowledge gaps and uncertainties beyond those directly associated with the OE industry are thus brought to bear on the EIA process.

The state of knowledge regarding device interactions is improving as the sector grows. The broad range of environmental interactions has largely been mapped out, though many areas of uncertainty remain.[19] Early research into the environmental interactions of OE technologies tentatively point towards a lack of significant direct negative impacts, but there are limiting factors that call for caution. Studies to date have typically covered a highly localised area or small-scale, as opposed to commercial, deployments and often rely on probabilistic models that require validation. They also suffer from uncertainties surrounding effect interactions and cumulative effects. There remains considerable uncertainty, particularly from the perspective of a regulator that requires a high level of certainty for decision-making and compliance purposes.

This lack of information affects OE projects in several ways. As with other maritime projects, the OE sector suffers from the general lack of established baseline data regarding the receiving environment. However, the OE sector differs in that the novel effects associated with devices, for example collisions with subsurface turbines, are challenging conventional baseline data collection methodologies and strategies. Further, impact assessments often cannot draw on an existing body of research to translate baseline findings into impact assessments. Finally, the diversity of device designs adds to the complexity of gathering data on the impact of OE devices. The potential effects on ecosystem components may vary between technologies and device types, making generalisations and extrapolations risky.

This scientific uncertainty is compounded by underdeveloped regulatory frameworks and EIA processes that have not been adapted to better support emerging technologies.[20] While some aspects of the relevant regulatory frameworks are being improved to better fit the OE context,[21] recent commentary suggests that OE bears a considerable regulatory burden, particularly in relation to EIA. For example, Merry (2014) notes that OE "attracts a depth of scrutiny from environmental regulators and statutory nature conservation bodies that more established marine industries such as fishing and shipping have managed to escape".[22] Nonetheless, as implementation of more integrated marine management approaches continues, maritime industries in general are increasingly held accountable for the impact of their activities. This is likely to increase as methodologies such as cumulative impact assessment become more firmly embedded in management practices.

Overall, there is a shortage of relevant, reliable public information regarding OE, resulting in an additional burden on developers who have to generate data on which to base environmental assessments. The resulting time and cost are a considerable barrier to the development of OE, while any privately collected data is less likely to be publicly available to benefit the sector as a whole.[23] This has a significant impact on the development and sustainability of OE.[24]

2.2.2 The EIA process for ocean energy projects

The process for EIA in the OE context essentially follows the outline given earlier in the chapter. The European Marine Energy Centre (EMEC)[25] has developed a set of guidelines, intended to encourage and assist developers to consider the full range and scale of impacts, which provides a framework of the EIA process for OE projects and highlights the scientific data and information that has to be generated.[26] The precise regulatory requirements vary by jurisdiction and will change as legislation is developed and implemented, but the guidelines provide a useful starting point.

As part of the screening and scoping stages, developers will have to identify available information regarding the proposed deployment site, while at stage 3 they will likely have to undertake baseline studies in order to describe the nature of the site.[27] This data is then used to identify potential known impacts and consequently cause–effect relationships to establish which components of the ecosystem require further consideration, evaluate the significance of predicted effects on those components, and identify potential mitigation options (stage 4) before compiling the documents for submission to the relevant regulatory authorities.

As the sector develops, the focus of the EIA process will likely shift to reflect emerging challenges and changing priorities. In the early stages of development, environmental interactions were primarily assessed in relation to individual devices. As the sector grows to include more expansive test sites, demonstration projects and eventually commercial deployment sites comprising different technologies, the assessment of potential cumulative effects will become progressively more important, as the scale and number of OE sites increase to warrant scrutiny against emerging marine management objectives.

2.3 Case studies

The following two case studies provide a practical perspective on how the EIA process can unfold in relation to OE projects, as well as highlight how the slow development of both science and regulation may affect the emerging OE industry.

2.3.1 Marine current turbines, Northern Ireland

Since April 2008, Marine Current Turbines' (MCT) SeaGen device has been deployed in Strangford Lough, Northern Ireland, and has generated over 3 GW of electricity to date. As the first deployment of a commercial-scale tidal energy device, MCT is a natural case study choice for any aspect of OE regulation and has already informed policy development in Northern Ireland and the UK.[28]

Over several years, MCT proceeded through each of the preceding stages of the EIA process, producing a Scoping Report, an Environmental Statement, an EIA report, an Environmental Action and Safety Management Plan, and an Environmental Monitoring Programme (EMP). MCT's approach to EIA represents a combination of the two concepts underpinning EIA as previously discussed. On the one hand, MCT established a scientific working group to advise on the scientific and operational details of the EMP and mitigation measures, while a broader liaison group was established to place the scientific research in context.

The EMP includes a comprehensive compliance and monitoring regime, which involves a range of pre- and post-deployment monitoring activities, including surveillance for and post-mortem evaluation of animal carcasses, establishment of a sonar marine mammal detection system to detect marine life moving past the turbine and to enable automatic precautionary shutdown; manual shutdown when a mammal is within 50 m of the device, land-based visual surveys, and telemetry studies and acoustic data logging for harbour porpoise activity.[29] This regime cost MCT £3 million over three years,[30] though the improved scientific knowledge gained through the EMP process allowed MCT to take an adaptive management approach, thereby reducing the company's mitigation burden in subsequent variations of MCT's SeaGen license.[31]

Although MCT agrees that the EMP process has ensured that deployment of their OE device is sustainable and environmentally friendly, it notes that the cost of undertaking such an intensive EIA process is burdensome and "could prove too much for future projects".[32] This onerous process was partly due to the fact that the device was deployed in an area already subject to a number of protective designations[33] and partly because it was the first deployment of its kind, necessitating extensive EIA supporting studies to convince regulators that the device would have minimal impact.

2.3.2 Case study: Crest Energy, New Zealand

In July 2006, Crest Energy applied to the local council for a range of consents to establish an array of 200 turbines on the seabed of Kaipara Harbour. The Crest Energy project underwent a lengthy process in the New Zealand court system in order to obtain consent and provides an example of how precaution-minded regulators may approach OE projects.[34]

A range of existing legislative and regulatory instruments exist in New Zealand that may apply to OE projects;[35] Crest Energy was the first to attempt to navigate the regulatory process. Crest followed the usual EIA process; however, a number of issues were identified regarding potential environmental impacts, resulting in two hearings in the Environmental Court. These issues included sustainable management of the OE resource, coastal planning, impacts on marine life, impacts on fish and fisheries, and the need for a suitable monitoring regime.

In the Court's view, the effect of the project on marine fauna and fisheries was a particularly important feature of the case. Having heard extensive evidence, the Court remained concerned about the infancy of the science surrounding environmental impacts but ultimately granted consent, subject to[36] a staged deployment of devices according to a predefined adaptive management plan; further caucusing of the experts, noise monitoring, a baseline monitoring regime for a period of two years prior to deployment, and more detailed research into the Maui's dolphin, an especially threatened species in the area.

Similar to MCT, Crest Energy's experience was ultimately positive, suggesting that decision makers will not reject projects simply because they are novel and scientific evidence is not yet clear. However, it is also clear that Crest bore an unusually high burden for being the first company in its jurisdiction to pursue a commercial-scale OE project because of the lack of a precedent or existing scientific knowledge concerning the potential environmental impacts.

2.4 Towards improved EIA processes

To date, a significant proportion of data and information collected on the effects of devices on the receiving marine environment have been collected through academic research projects, often funded through national research budgets or via intergovernmental organizations, in particular the EU. This type of research is independent of the EIA process but necessary to build fundamental scientific understanding of marine ecosystems, their functioning and the species and habitats therein. Such funding, however, is being cut and does not provide a sustainable mechanism to ensure good scientific information for important decisions relating to further marine development. EIAs provide much needed validation of modelled or theoretical studies, and the management and policy domain needs to be responsive to such new knowledge in order to refine consenting processes.

Given the likely importance of EIA processes, the preceding case studies highlight a number of issues with the application of unaltered EIA processes to OE projects:

1   The lack of scientific certainty regarding impacts causes the EIA process to be time-consuming. The implementation of MCT's EMP began in June 2005 and ended in 2011 with a final report concluding that no major impacts on marine mammals were detected in three years of post-installation monitoring. Crest Energy made its first application for permits in July 2006 and received them in March 2011, with ongoing scientific studies to be conducted once deployment commences.
2   EIA approvals came at a considerable cost to the developers. The reporting and research requirements for EIA are expensive, with post-installation monitoring adding further cost. This is in addition to the opportunity cost incurred due to a delay in consenting and deployment.
3   The availability of suitable scientific data regarding environmental impacts will be a key factor in the regulatory process (in particular the establishment of adequate baseline data that enables the likely effects of a development to be identified and distinguished from baseline trends).[37]

4   There are issues with regulator capacity. Regulators require funding and access to tools that enable the acquisition of scientific data independent of the EIA process, particularly in light of integrated marine management ambitions. For example, to reduce regulator uncertainty during consenting, regulators would benefit from fit-for-purpose baseline data and from threshold data for the broader environment in order to provide much-needed context for EIAs, which are limited in spatial and temporal scope. Regulatory capacity is also critical to enforce compliance with legal instruments, including permits and licenses awarded to developers and conditions that may be attached to licenses. Without the appropriate resources and capabilities, there is a risk that the onus falls on developers, adding to the financial risk of development.

5   Innovation may be hampered if EIA processes continue to require OE proponents to bear onerous responsibilities for generating new scientific information, proving new technology and providing examples to regulators. The final SeaGen EMP report echoes this sentiment, stating that "the SeaGen EMP provides an ambitious plan beyond what might be expected of future projects now that more knowledge is available".[38]

These early EIA processes are crucial because minimising environmental damage and maximising environmental benefits will inevitably attract investors and reassure governments and decision makers who may otherwise view potential environmental impacts as a barrier.[39] This suggests that each successful and environmentally acceptable deployment generates useful knowledge that can be factored into the EIA process and wider policy environment for later projects. As such, a key question is how EIA regimes can encourage the generation and dissemination of knowledge, including methodologies and impact assessment outcomes that will provide sufficient environmental and scientific certainty for future projects. Such knowledge sharing would encourage the emergence of consistent and incrementally improved EIAs, counter to the common trend in EIA practice of diverse procedures that hinder regulatory review and jeopardise cost effectiveness.[40]

However, the consenting of one OE project will not necessarily mean that regulators will be more accepting of other projects or that OE EIA will become less onerous. Continued regulatory uncertainty, the scale of a proposed project and highly localised environmental conditions may still necessitate an extensive EIA process and the development of a comprehensive and detailed EMP. Until sufficient OE projects are deployed in enough locations covering a variety of habitats and are deployed at scales where the likely cumulative effects are better understood, uncertainties resulting in delays during the EIA process seem set to continue, highlighting the Catch-22 scenario facing the OE industry.

In this context, tailoring the EIA process to OE regulation will be a difficult balancing act, weighing environmental interactions, on the one hand, with the need to develop an innovative source of clean energy, on the other.[41] Two core aspects to the regulation of the environmental interactions of OE must be discussed: the project-level regulatory framework, exemplified by the EIA processes previously, discussed, and the broader marine governance framework within which projects are consented.

2.5 Project-level regulation

2.5.1 Precaution versus risk

In the context of imperfect information, two general approaches can be taken that will set the tone for EA processes: the precautionary approach and a risk-based approach. The choice has the potential to shape regulation and facilitate or hinder industry development, and as such it is a heated issue (e.g. offshore wind farm developers and regulators in the UK have had a "fiercely contested" debate over which model is appropriate).[42] This lack of consensus also extends to the highest legislative levels, with policy drivers behind emerging EU marine environmental legislation underpinned by different interpretations of sustainability, further complicating the regulatory landscape.[43]

The precautionary principle is a well-entrenched tenet of environmental law that dictates that if an action risks causing harm to the environment, in the absence of scientific consensus or certainty that the action is not harmful, the burden of proof falls on the proponent of the action to provide this certainty.[44] Within the EU, the European Commission recognises that "finding the correct balance so that proportionate, non-discriminatory, transparent and coherent actions can be taken, requires a structured decision-making process with detailed scientific and other objective information".[45] Regulators have traditionally applied a strict interpretation of the precautionary approach, including towards renewable energy projects, resulting in developers having to meet stringent levels of certainty that significant impacts will not result, despite there being no evidence, as yet, to prove significant direct environmental impacts as a result of deployment of individual devices or small arrays.[46] Zealous and arguably disproportionate interpretations of the precautionary principle by judicial institutions likely contribute to the cautious approach to risk displayed by regulators.[47]

The nature of the marine environment, coupled with the challenges of identifying and quantifying the apparently low significance of OE effects, leaves developers with the difficult task of estimating a "relatively small change in a highly variable environment" in order to prove to the regulator that there are no significant impacts.[48] By failing to allow for some risk, a rigidly applied precautionary approach may result in a potential impact being perceived as being more likely to occur than it truly is and thereby "threatens to be paralyzing".[49] It might also result in more mitigation and more compensation than may be necessary owing to the potential scale and likelihood of the impacts associated with the uncertainties. A strict application of the precautionary principle does not actively seek to reduce scientific uncertainty and therefore does not have a goal of improving decision making over time by reducing the uncertainties.

By contrast, a risk-based approach seeks to inform decision making through an understanding of the scientific uncertainties and associated consequences in terms of likelihood and magnitude of potential impact.[50] Risk-based regulation can therefore provide a systematic framework that prioritises regulatory activities and the scientific studies required to meet regulatory requirements, according to an evidence-based assessment of risk.[51]

2.5.2 Adaptive management

Adaptive management (AM) is a process-oriented and risk-based approach to regulation, with a long history in the fields of conservation and resource management.[52] The US Department of the Interior defines adaptive management as a "decision process that promotes flexible decision making that can be adjusted in the face of uncertainties as outcomes from management actions and other events become better understood".[53] Adaptive management requires that all parties work together to share existing knowledge, identify scientific uncertainties, fill knowledge gaps and consider multiple mitigation options. Deployment can then be permitted under a monitoring regime, and the regulatory approach and requirements can be adapted on the basis of newly obtained information.

Adaptive management acknowledges that scientific understanding of an ecosystem will always be incomplete and allows for management actions to be readjusted over time to take new scientific data and knowledge developed into account. Adaptive management thereby treats regulatory decisions and deployments as experiments that can be used not only to change deployment plans but also to learn about the interaction with the receiving environment.[54] This can help regulators to better understand and account for uncertainties by enabling each management option to be evaluated and continually reviewed as learning occurs through monitoring.[55]

Monitoring is a critical part of such an approach. In normal practice, environmental monitoring is generally required as a term or condition of one or more of the granted consents. The initial set-up phase will involve deciding on specific management objectives, which might include, for example, the acceptable number of mortalities of a species, the acceptable level of displacement of marine mammals, the maintenance of noise at an acceptable level, and so on. Specific monitoring requirements should then be agreed upon between the regulator and the developer.[56] The data and information generated from this monitoring subsequently is used to adapt decision making where necessary, based on what has been learned from the previous management actions.

Adaptive management can be used where EIA cannot confidently provide accurate predictions. In this way, the EIA process can itself be an integral part of generating scientific information and knowledge, assuming the resultant information is made available and subsequent monitoring programmes are sufficiently well designed to meaningfully inform regulators. By responding to and managing feedbacks observed during deployment, instead of blocking them out at the start of the consenting process through unduly restrictive conditions, adaptive management has the potential to "avoid the pathology of natural resource management that threatens the existence of many social and economic activities".[57]

Within the EU, the revised EIA Directive[58] makes post-project approval monitoring a requirement,[59] obliging Member States to determine the procedures for the monitoring of "significant adverse effects" on the environment. The parameters to be monitored and the duration of the monitoring "shall be proportionate to the nature, location and size of the project and the significance of its effects on the environment". In essence, this means that Member States must monitor the implementation of projects that have been subject to an EIA. The practical implications of this include having measures in place to ensure that the developer complies with any conditions attached to the consent concerning mitigation, compensation measures and the like and monitoring to identify any significant adverse effects that might arise, including any unanticipated or unforeseen significant effects that might surface during implementation. In theory, the inclusion of monitoring in the revised EIA Directive should facilitate adaptive management by clearly acknowledging uncertainty and proposing practical monitoring arrangements to verify the correctness of the predictions made during the EIA process. If implemented correctly, this should result in a move away from the static/reactive approach of EIAs based on predictions and a single-response model towards a more flexible EIA.

A key challenge presented by embedding an AM approach to consenting is the time frames involved. The purpose of an AM approach is to reduce scientific uncertainty which, by its very nature, cannot happen quickly. In practice, this means that a regulator will have to decide whether it is going to continue making decisions based on the existing level of knowledge or wishes to take a more progressive approach and reduce scientific uncertainty through embedding AM principles in its processes and procedures. This could result in additional costs for developers, but it is probable that developers will be more inclined to bear the expense of monitoring within an AM framework if cost-effective consenting processes are in force and if 'savings' have been realised at earlier stages, such as during site characterisation surveys.

2.5.3 Survey, deploy and monitor

The survey, deploy and monitor (SDM) approach is an example of an adaptive management approach that is used in the pre-consenting period. It was first introduced by Marine Scotland, the Scottish body responsible for marine planning and consenting, and is embodied in Scotland's *Survey, Deploy and Monitor Licensing Policy Guidance*.[60] SDM informs site characterisation survey requirements by enabling EIA requirements to be adjusted at the scoping stage, in cooperation with the regulator, to reduce the burden of collecting baseline data to inform EIAs on small-scale projects or projects of low environmental risk.

Under the SDM approach, the duration of site characterisation surveys and the level of monitoring are then determined by the overall risk profile of the project. The risk profile is calculated on the basis of the environmental sensitivities of the area, the scale of development, and the specificities of device risks.[61] On the basis of the calculation, a project proposal is categorised as low, medium or high risk. A proposal classified as high risk requires a minimum of two years of site characterisation data. The same is true for medium-risk proposals, but Marine Scotland has the discretion to relax monitoring requirements in cases where they feel this is warranted. For example, a small-scale project in an area of low environmental sensitivity may only need one year of site characterisation data to inform the consenting process. In certain cases, this can be further reduced, dependent on the specific site and level of existing data.

This SDM approach, therefore, exercises a certain level of precaution by utilising existing scientific information to exclude certain sites, while also allowing deployment in other areas, thereby enabling additional studies to be undertaken. This will then provide scientific data in relation to both the technology and the receiving environment. Decision making based on this type of approach is therefore risk based and proportionate to the risk profile of the development proposed.[62]

By providing for an evidence-based decision on consenting applications, the SDM approach ensures that existing data is fully utilised before requiring additional studies. This should allow more rapid deployment and therefore provide developers with the opportunity to generate new scientific information regarding effects and impacts, without having to study every potential impact in advance to a high level of certainty. Furthermore, where the characterisation alerts the regulator to the need for further information regarding a particular aspect of the receiving environment or project, the SDM approach allows the EIA process to continue. While the outcome of the consent application may not be determined until all additional requirements have been met, the SDM approach ensures that any additional requests for data or studies is proportionate to the development being proposed.

Overall, this process attempts to apply a more appropriate, proportionate and individualised approach to licensing and EIA based upon the nature and circumstances surrounding the development proposal, the level of scientific information already available and the need to improve knowledge rather than a one-size-fits-all approach.

2.5.4 The Rochdale Envelope

Whereas the approaches discussed already primarily centre on the receiving environment, the Rochdale Envelope focuses attention on the generation of new scientific information in relation to the technology itself (which can, in turn, feed into other processes, such as the classification of risk for the purposes of a SDM policy). The Rochdale Envelope approach, named after a UK planning law case,[63] allows a project description to be broadly defined, within a number of agreed parameters, for the purposes of a consent application. This allows for a level of flexibility while a project is in the early stages of development. As development progresses, further scientific study takes place, creating more confidence regarding impacts that can feed into the ultimate project design.

The Rochdale Envelope should enable projects to take full advantage of the constant evolution of technologies, rather than locking in a detailed proposal at the consent application stage. This is essential to the planning process at this early stage of OE industry development as it enables scientific advancements to be taken into account during the installation rather than in the earlier design phases.

Despite proving popular with offshore wind and OE developers, there is little literature available on best practice use of the Rochdale Envelope approach. The UK's Infrastructure Planning Commission has issued some limited guidance for developers.[64] Expanded research and guidance would be welcome, as the incorporation of the Rochdale Envelope approach into EIAs for offshore wind has demonstrated some of the potential pitfalls of the approach. In particular, EIAs for offshore wind have commonly overstated the likely upper limit of the envelope, which has caused problems where multiple projects in one are assessed as having a much higher risk of causing cumulative impacts than is truly likely.[65]

3 Strategic environmental assessment

Strategic environmental assessment (SEA) is a systematic decision support process, aimed at ensuring that environmental considerations are fully accounted for in plan and programme making at a strategic level. SEA provides an additional overarching layer of EA and could play a role in reducing the project-level EIA burden by ensuring a proactive approach to environmental data and impacts at an earlier stage in the development of an activity or industry.[66] SEA is undertaken either independently or in the course of a wider MSP process (see later in the chapter). The early conception of SEA was as an extension of project-level EIA, centred on the production of a written report. More recent formulations, now considered good practice, instead recognise SEA as an iterative, proactive and integrated process for evaluating the environmental impacts of a strategic programme or plan.

SEA can potentially strengthen and streamline EIA at the project level by identifying the scope of potential impacts and information needs, addressing strategic issues and concerns related to the justification of proposals and reducing the time and effort necessary to conduct individual assessments. In theory, if a SEA can contribute baseline data and/or assist with site selection, developers will have to spend less time and capital developing detailed EIAs and forging a process for project approval, and regulators will have recourse to the broader context necessary to consider project-level impacts. SEAs also have the potential to disclose licensing and consenting requirements, particularly helpful in jurisdictions that have not yet streamlined these processes or consolidated them.

SEAs for OE have been conducted in Scotland and Nova Scotia, Canada (discussed later in the chapter), as well as Ireland[67] and Northern Ireland.[68]

In England, an offshore SEA has been conducted on all forms of energy development, renewable and non-renewable,[69] while Spain has conducted an SEA for offshore wind only.[70]

3.1 SEA in Scotland

In 2007, the Scottish Government commissioned an SEA to examine the environmental interactions of OE, inform a strategy for OE development and provide a reference for regulators and stakeholders to support project-level decision making.[71] The SEA covered the west and north coasts of Scotland from Shetland to the Solway Firth to a distance of 12 nm offshore. This area was then subdivided into eight separate areas, based on where the main wave and tidal resources are located, following consultation with developers.[72] The information generated from the SEA process was subsequently utilised in the development of a series of more detailed Regional Locational Guidance on various types of technologies.[73]

The complexity of the environment, large knowledge gaps and limited understanding of the environmental interactions of OE technologies necessitated a different approach to the SEA process than taken with 'traditional' SEAs on land. The SEA was essentially a large-scale environmental baseline collation exercise used to provide an indication of the environmental constraints upon sections of the marine environment. The SEA, therefore, contained more detail than typical SEAs. Taking the high-level approach typically used in SEAs or reverting to a more project-centred approach would have led to some potentially significant impacts being overlooked. The SEA therefore provided the Scottish Executive and Marine Scotland, as decision makers on OE developments, with an invaluable resource for ensuring that decisions about OE are taken in an informed manner.[74]

The Scottish SEA was productive insofar as it identified key areas of sensitivity and resulted in the production of a number of relevant studies. However, although all parties agreed that data gaps were a key shortcoming and that there was an obvious need to develop a structured and prioritised research programme, the SEA did not contain any recommendations or provisions for filling the knowledge gaps identified. The Scottish Government made subsequent efforts to this effect. Shortly after the completion of the SEA, a follow-on group was formed to identify the projects that needed to be initiated and funded to work towards filling knowledge gaps. The Programme was established in 2011 to provide scientific input and support to policymaking and licensing of OE,[75] and its Scottish Marine Renewables Research Group has completed 15 scientific research projects to date, with a further 19 ongoing and additional projects planned for the future.[76]

3.2 SEA in Nova Scotia, Canada

In 2005, OE developers, seeking to deploy OE technology in the highly prospective Bay of Fundy area in Nova Scotia (NS), approached the Nova Scotia Department of Energy. The Government responded by requesting the Ocean Energy and Environment Research Association (OEER) to undertake a SEA process to inform the development of the resource. As with the Scottish SEA, the objective was to assess the social, economic and environmental effects of OE deployment in the region in order to inform decisions regarding the OE industry.[77]

From an environmental perspective, the key outcome of the SEA was the conclusion that development should proceed incrementally, with the identified key knowledge gaps addressed at each step before further deployment decisions were made. As with OE industries generally, scientific uncertainties prevented confident predictions of ecological effects of (1) specific devices, (2) OE array infrastructure, and (3) the cumulative effects of OE infrastructure and other maritime activities.[78] A high level of agreement was reached on key issues, but time and resource limitations precluded deeper agreement on substantive issues.[79] A further limitation caused by the small level of funding available is that knowledge gaps were identified but generally not filled as part of the SEA process. The NS SEA did, however, demonstrate that considerable new knowledge can be generated and shared through mutual learning. During the process, developers provided insight into ideal conditions for the development of a variety of technologies, while the fishing industry provided information on local conditions.

3.3 Improving scientific knowledge through SEA: lessons learned

Though the experience with SEA in the OE context is limited, the two preceding examples provide some lessons learned in terms of improving scientific knowledge. Though it may be some time before the impacts of the SEAs are known, the role of SEAs in improving the science is clearly much stronger in the Scottish SEA. The NS SEA focused much more on high-level identification of impacts, without also providing sufficient detail at the lower levels to benefit developers and encourage specific additional research. As a result, the SEA's recommendations are general in nature. In contrast, the Scottish SEA was more explicitly undertaken to identify knowledge gaps and to act as part of the foundation for a much broader planning framework for OE. As such, the Scottish SEA enters into a much higher level of detail. A similar approach was followed in Ireland with the creation of an Offshore Renewable Energy Steering Group and Environment Working Group who have coordinated the production of guidance on Environment Impact Statements and Natura Impact Statements and on Marine Baseline assessments and Monitoring Activities for Offshore Renewable Energy projects.[80]

It was noted that in NS the SEA process itself was effective at bringing to light existing research and knowledge; the process was conducted as a roundtable, inviting various stakeholder groups to participate. This process brought to light a wide range of existing scientific knowledge and data but highlighted that this was often sector specific, unpublished, or drawn from practical experience rather than empirical study. This knowledge, though potentially valuable, has therefore not yet been part of the broader discussion.

A further lesson learned is the need for follow-up actions to give effect to the SEA and its recommendations. The Scottish Government has given lasting influence and effect to the SEA by initiating a comprehensive research programme to fill in the knowledge gaps identified. By contrast, the NS Government has not since followed the SEA with additional research, though this may of course change as the industry further develops in this region.

4 Appropriate assessment

The EU Habitats Directive[81] aims to conserve rare, threatened or endemic animal and plant species. Member States must designate Special Areas of Conservation (SACs) to ensure the favourable conservation status of each habitat type and species throughout their range in the EU. This is complemented by the Birds Directive,[82] which provides for the designation of Special Protection Areas (SPAs) for particularly threatened species and all migratory bird species. Together, SACs and SPAs form the Natura 2000 network of protected sites. Member States must take appropriate conservation measures to maintain and restore the habitats and species for which the site has been designated to a favourable conservation status,[83] and ensure that damaging activities which could significantly disturb species or deteriorate the habitats of the protected species or habitat types are avoided.[84]

Any plan or project not directly connected with the site but likely to have a significant effect on it, either individually or in combination with other plans or projects, is subject to appropriate assessment (AA) of its implications for the site in view of the site's conservation objectives. Competent authorities can approve the plan or project only after having determined that it will not adversely affect the integrity of the site concerned. This is the key difference between AA and other forms of EA: if the AA is negative and, despite mitigation measures, impacts on integrity are still anticipated, then the plan or project can be authorised only if the plan or project is justified by "imperative reasons of overriding public interest".[85] If the project does not meet these criteria, the project cannot be authorised.

The EC has issued guidance in relation to Natura 2000 sites and various sectors, including wind energy.[86] The Commission has also clarified that inherent scientific uncertainties need not preclude an assessment of no impact to integrity by using a flexible approach, stating that "minor remaining uncertainties should however not block or restrain projects indefinitely".[87] Furthermore it states that the nature of remaining uncertainties should be managed "through targeted monitoring and adaptive strategies", thereby reflecting the principles of adaptive management.[88] The guidance referred to relates to estuaries and coastal zones, specifically to port development and dredging in those areas, so the extent to which the guidance can apply to OE is unclear.

5 Marine Spatial Planning (MSP)

Marine spatial planning (MSP), discussed in Chapter 6, is a relatively recent approach to planning and managing the marine environment and marine activities. MSP seeks to integrate various maritime sectors and should be eco-system based, participatory, strategic, adaptive and tailored to suit the needs of a predetermined marine area. The aspiration is that MSP will provide a stable and transparent planning system for maritime activities and users within agreed environmental limits, so as to ensure that marine ecosystems and their biodiversity remain healthy.

Maritime activities are traditionally managed on a sector-by-sector basis. This limits how one activity interacts with another activity occurring in the same place, and impact assessment tends to be limited to that one activity as opposed to assessing the wider cumulative effects of that activity on the receiving environment. MSP, therefore, has huge potential to influence maritime activities now and into the future and will become an important enabling framework for OE as its implementation becomes more widespread. If planning proceeds in a sectoral manner, the likelihood of conflicts between sectors and activities increases.

Work conducted as part of the IEA's Ocean Energy Systems Annex IV on the Environmental Effects of Marine Renewable Energy, found that, of the 11 countries that responded, four have formal MSP processes in place, three do not and four do not have MSP *per se* but do have coastal management plans in place that can include marine and coastal uses such as shipping, fisheries and conservation.[89] As MSP is one of the more recent approaches to marine environmental management and OE is a comparatively new use of the same environment, it is difficult to draw firm conclusions about the extent to which one is influencing the other currently (see Chapter 6). There is a strong desire for more integrated planning, and implementation of MSP should go some way towards solving certain persisting issues associated with development at sea. Consenting systems will have to function within the wider MSP framework in future.

6 Conclusion

The experience of the OE industry with existing EA frameworks suggest that substantial reform is required if economic, social and environmental goals for new marine industries like OE are to be successfully balanced. It is essential that innovative new marine-based technologies are supported by consistent, transparent and complementary regulatory frameworks, as well as strong scientific evidence regarding environmental interactions. EA, in its various forms, can play a part in providing this. Firstly, EA processes must do more to incorporate all available information into regulatory decisions to the fullest extent possible. Secondly, EA and broader marine governance initiatives must be a conduit for the dissemination of knowledge generated during their respective processes. In this regard, both site-specific EIAs, which can provide high-resolution data much needed by regulators, and SEAs, which can provide the context needed to interpret identified EIA impacts, should be highly complementary. This requires much greater integration between the processes and their competent authorities than is currently observed. Thirdly, marine governance frameworks must complement EA processes by identifying knowledge gaps and by providing for the advancement of scientific knowledge.

An adaptive management approach is clearly needed in order to make the best use of existing scientific data and to permit deployment of devices that can generate new information and knowledge. The Scottish SDM policy provides one model for factoring in a certain level of risk in project-level decision making, as does experience with the Rochdale Envelope approach. EA in its current guise is often concerned with maintaining the initial state or baseline conditions rather than providing a strategic plan aimed at reducing uncertainties and mitigating environmental impacts. It cannot, therefore, be considered as an adaptive management approach. Over time, the EA framework can transition to one based on adaptive management principles, providing a collaborative environment in which developers and regulators can progress scientific knowledge, improved baseline data and reduce the time and cost currently associated with such processes.

The experience in Scotland and Nova Scotia with SEA suggests that a detailed SEA process can help to identify knowledge gaps and provide a foundation for a risk-based approach; however, sufficient political will and funding must be available for the SEA process, and follow-up action must be taken to fill the identified knowledge gaps. Further, SEAs need to address the EIA-based constraints and become more than an "EIA-plus" approach, to better provide the regional picture required to enable integrated marine management.[90] Marine spatial planning should result in a more coordinated and coherent approach to marine planning and management, but this will be wholly dependent on how it is implemented at national, regional and local levels. Adaptive management approaches also support the implementation of MSP, as well as the environmental objectives associated with the EU's Marine Strategy Framework Directive.

This chapter has necessarily provided only a brief overview, and further research is required to improve marine governance structures, especially where the new technologies or industries are likely to have relatively low environmental impacts but where apparently minor, incremental effects may have a significant cumulative effect on a degraded ecosystem. In particular, research is needed into the practical effects of EIA regimes on the sustainable development of OE and other technologies. Further research is also required regarding jurisdictions that have begun to use some of the management approaches discussed in this chapter to assess whether they are having the desired impact, what are the best practices in implementing them and how other jurisdictions might successfully follow these examples. With further research, reform and experience with implementation, EIA processes can contribute to emerging marine governance frameworks, advance scientific knowledge and help progress OE development.

Notes

1 For general background, see e.g. Canadian Environmental Assessment Agency (1996) *Environmental Assessment in a Changing World: Evaluating Practice to Improve Performance*. International Study of the Effectiveness of Environmental Assessment, prepared by B. Sadler. Ottawa: Canadian Environmental Assessment Agency in

cooperation with the International Association for Impact Assessment, p. 263; Morgan, R. K. (2012) Environmental Impact Assessment: The State of the Art. *Impact Assessment and Project Appraisal*, 30(1), pp. 5–14.

2 See U.S. National Environmental Policy Act 1969 (NEPA). Available at: www.fws. gov/r9esnepa/RelatedLegislativeAuthorities/nepa1969.PDF

3 Caldwell, L. K. (1981) Environmental Impact Analysis (EIA): Origins, Evolution, and Future Directions. *Review of Policy Research*, 8(1), pp. 75–83.

4 Petts, J. (1999) Introduction to Environmental Impact Assessment in Practice: Fulfilling Potential or Wasted Opportunity? In: Petts, J. (ed.) *Handbook of Environmental Impact Assessment: Volume 2: Environmental Impact Assessment in Practice*. Oxford: Blackwell Science, pp. 3–9.

5 Consideration of alternatives is a consistent weakness in the EIA process; see European Communities (2009) *Report from the Commission to the Council, the European Parliament, the European Economic and Social Committee and the Committee of the Regions on the Application and Effectiveness of the EIA Directive (Directive 85/337/EEC, as Amended by Directives 97/11/EC and 2003/35/EC)*, COM(2009) 378 Final. Brussels: Commission of the European Communities. Available at: http://eur-lex.europa.eu/ legal-content/EN/TXT/?uri=CELEX:52009DC0378; and also Pope, J., et al. (2013) Advancing the Theory and Practice of Impact Assessment: Setting the Research Agenda. *Environmental Impact Assessment Review*, 41, pp. 1–9.

6 There may also be a formal process for engaging with consultees.

7 These conditions will generally include monitoring and compliance, which could alternatively be viewed as separate steps in the process.

8 There is a rich literature on this interaction between law and technology. For an early discussion of a new marine industry, see Nyhart, J. D. (1974) The Interplay of Law and Technology in Deep Seabed Mining Issues. *Virginia Journal of International Law*, 15(4), pp. 827–868.

9 Cashmore, M. (2004) The Role of Science in Environmental Impact Assessment: Process and Procedure Versus Purpose in the Development of Theory. *Environmental Impact Assessment Review*, 24(4), pp. 403–426.

10 Lee, B., Haworth, L., and Brunk, C. (1995) Values and Science in Impact Assessment. *Environments*, 23(1), pp. 93–100.

11 Cashmore, M. (2004) n. 9.

12 Lawrence, D. P. (1997) The Need for EIA Theory-Building. *Environmental Impact Assessment Review*, 17(2), pp. 79–107.

13 For general discussion on the role of science in EIA, see e.g. n 9. For empirical studies on the scientific content of EIA documents in practice, see e.g. Culhane, P. J., Friesema, H. P., and Beecher, J. A. (1987) *Forecasts and Environmental Decision-making: The Content and Predictive Accuracy of Environmental Impact Statements*. Boulder, CO: Westview Press; Malik, M., and Bartlett, R. V. (1993) Formal Guidance for the Use of Science in EIA: Analysis of Agency Procedures for Implementing NEPA. *The Environmental Professional*, 15(1), pp. 34–45.

14 Cashmore, M. (2004) n. 9.

15 Ibid. p. 410.

16 For example, regulators can draw on over a century of experience with conventional hydropower technologies.

17 Smith, A. K. (2008) Impact Assessment in the Marine Environment – The Most Challenging of All. In: *Impact Assessment in the Marine Environment*. Perth: International Association of Impact Assessment Annual Conference, p. 5. Available at: http://conferences.iaia.org/2008/pdf/themeforums/TF4-1_sector_marine_Smith.pdf

18 Borja, Á., et al. (2016) Overview of Integrative Assessment of Marine Systems: The Ecosystem Approach in Practice. *Frontiers in Marine Science*, 3(55), p. 68; Elliott, M., et al. (2015) *Force Majeure*: Will Climate Change Affect Our Ability to Attain

Good Environmental Status for Marine Biodiversity? *Marine Pollution Bulletin*, 95(1), pp. 7–27.

19 See Boehlert, G., and Gill, A. (2008) Environmental and Ecological Effects of Ocean Renewable Energy: A Current Synthesis. *Oceanography*, 23(2), pp. 68–81; Inger, R., et al. (2009) Marine Renewable Energy: Potential Benefits to Biodiversity? An Urgent Call for Research. *Journal of Applied Ecology*, 46(6), pp. 1145–1153; Simmonds, M. P., et al. (2010) Marine Renewable Energy Developments: Benefits Versus Concerns. Chippenham, UK: Paper SC/62/E8 presented to the IWC Scientific Committee, 2010 (unpublished); Frid, C., et al. (2012) The Environmental Interactions of Tidal and Wave Energy Generation Devices. *Environmental Impact Assessment Review*, 32(1), pp. 133–139; Linley, A. (2012) Environmental Interactions with Marine Renewable Energy. *Marine Scientist*, 40, pp. 22–25; Hammar, L., et al. (2015) A Probabilistic Model for Hydrokinetic Turbine Collision Risks: Exploring Impacts on Fish. *PLoS ONE*, 10(3), pp. 1–25; Uihlein, A., and Magagna, D. (2016) Wave and Tidal Current Energy – A Review of the Current State of Research Beyond Technology. *Renewable and Sustainable Energy Reviews*, 58, pp. 1070–1081.

20 Huertas-Olivares, C. and Norris, J. (2008) Environmental Impact Assessment In: Cruz, J. (ed.) *Ocean Wave Energy: Current Status and Future Perspectives, Green Energy and Technology*. Berlin: Springer, pp. 397–423.

21 O'Hagan, A. M. (2012) A Review of International Consenting Regimes for Marine Renewables: Are We Moving Towards Better Practice? In: *Fourth International Conference on Ocean Energy*. Dublin: Ocean Energy Systems; Wright, G. (2014) Regulating Marine Renewable Energy Development: A Preliminary Assessment of UK Permitting Processes. *Underwater Technology: The International Journal of the Society for Underwater*, 32(1), pp. 1–12.

22 Merry, S. (2014) Marine Renewable Energy: Could Environmental Concerns Kill Off an Environmentally Friendly Industry? *Underwater Technology: The International Journal of the Society for Underwater*, 32(1), pp. 1–2.

23 Sterne, J. K., et al. (2009) The Seven Principles of Ocean Renewable Energy: A Shared Vision and Call for Action. *Roger Williams University Law Review*, 14(3), pp. 600–623. Nonetheless, some efforts to collaborate on data sharing are taking place. See, e.g. Tethys. Available at: http://tethys.pnnl.gov/

24 Jeffrey, H., and Sedgwick, J. (2011) ORECCA *European Offshore Renewable Energy Roadmap*. Produced as part of the ORECCA Coordinated Action Project. Edinburgh: University of Edinburgh, p. 19. Available at: https://setis.ec.europa.eu/energy-research/sites/default/files/project/docs/2012-01-19%20241421%201097326%20ORECCA%20Roadmap%20ExecSummary_Sep2011.pdf

25 An OE test and research centre based in Orkney in Scotland.

26 European Marine Energy Centre (2005) *Environmental Impact Assessment (EIA) Guidance for Developers at the European Marine Energy Centre*. Stromness, Orkney: EMEC, p. 22. Available at: https://tethys.pnnl.gov/sites/default/files/publications/EMEC_2005.pdf

27 These baseline studies could relate to a range of issues, such as determining local fish and mammal populations. At EMEC, baseline data is already available for use in EIA.

28 Marine Current Turbines (MCT) (2011) *SeaGen Environmental Monitoring Programme: SeaGen Biannual Environmental Monitoring March 2010 – October 2010*. Report produced by Royal Haskoning. Edinburgh: Royal Haskoning, p. 81. Available at: www.marineturbines.com/sites/default/files/SeaGen-Environmental-Monitoring-Programme-Final-Report.pdf

29 Ibid.

30 Riddoch, L. (2009) Seal of Approval. *The Nature of Scotland*, (Winter Issue), pp. 22–23.

31 Ibid.

32 Riddoch, L. (2009) n. 30.
33 Designated as the Strangford Lough Special Protection Area under the Birds Directive and a Special Area of Conservation under the Habitats Directive, as well as national designations.
34 Wright, G. (2011) A Tidal Power Project. *New Zealand Law Journal*, (September), pp. 260–261.
35 Wright, G., and Leary, M. (2011) Marine Energy. *New Zealand Law Journal*, (August), pp. 227–230.
36 A marine mammal with a vanishingly small population of approximately 100 individuals, confined to a small geographic range, meaning that the death of one dolphin could threaten the survival of the entire species.
37 Willsteed, E., et al. (2016) Assessing the Cumulative Environmental Effects of Marine Renewable Energy Developments: Establishing Common Ground. *Science of The Total Environment*, 577, pp. 19–32.
38 Marine Current Turbines (MCT) (2011) *SeaGen Environmental Monitoring Programme: SeaGen Biannual Environmental Monitoring March 2010 – October 2010*. Report produced by Royal Haskoning. Edinburgh: Royal Haskoning, p. 81. Available at: www.marineturbines.com/sites/default/files/SeaGen-Environmental-Monitoring-Programme-Final-Report.pdf
39 *Annex I to the Protocol on Environmental Protection to the Antarctic Treaty: Environmental Impact Assessment*, pp. 397–398.
40 Pope, J., et al. (2013) Advancing the Theory and Practice of Impact Assessment: Setting the Research Agenda. *Environmental Impact Assessment Review*, 41, pp. 1–9.
41 United Nations General Assembly (2012) *Report on the Work of the United Nations Open-Ended Informal Consultative Process on Oceans and the Law of the Sea at Its Thirteenth Meeting A/67/120*. New York: United Nations, p. 14. Available at: http://undocs.org/A/67/120. OE "could foster increased energy security, generate employment and play a role in mitigating the impacts of climate change. At the same time, the importance of assessing and studying the impacts of [OE], including on the marine environment, was stressed by several delegations".
42 RenewableUK (2011) *Consenting Lessons Learned: An Offshore Wind Industry Review of Past Concerns, Lessons Learned and Future Challenges*. London: RenewableUK, p. 28.
43 Qiu, W., and Jones, P. J. S. (2013) The Emerging Policy Landscape for Marine Spatial Planning in Europe. *Marine Policy*, 39(1), pp. 182–190.
44 Cross, F. B. (1996) Paradoxical Perils of the Precautionary Principle. *Washington and Lee Law Review*, 53(3), pp. 851–925.
45 European Commission (2000) *Communication from the Commission on the Precautionary Principle*. Brussels: Commission of the European Communities. COM/2000/0001 final. Available at: http://eur-lex.europa.eu/legal-content/EN/TXT/PDF/?uri=CELEX:52000DC0001&from=EN
46 RenewableUK (2011) n. 42.
47 McIntyre, O. (2013) The Appropriate Assessment Process and the Concept of Ecological 'Integrity' in EU Nature Conservation Law. *Environmental Liability*, 21(6), pp. 203–215.
48 Ibid.
49 Sunstein, C. R. (2003) Beyond the Precautionary Principle. *University of Pennsylvania Law Review*, 151, pp. 1003–1058.
50 Le Lièvre, C., et al. (2016) *Legal Feasibility of Implementing a Risk-Based Approach to MRE Consenting and Compatibility with Natura 2000 Network: Deliverables 2.3 and 2.4, RiCORE Project*. Cork: UCC, p. 53.
51 Peterson, D., and Fensling, S. (2011) Risk-Based Regulation: Good Practice and Lessons for the Victorian Context. In: *Victorian Competition and Efficiency Commission Regulatory*

Conference. Melbourne: Victoria State Department, p. 33. Available at: www.parlia
ment.vic.gov.au/images/stories/committees/rrc/IFSR/Submissions/045.20120907.
Victorian_Government.pdf

52 Adaptive management is sometimes also called adaptive resource management in
these fields.

53 Williams, B. K., Szaro, R. C., and Shapiro, C. D. (2009) *Adaptive Management: The
U.S. Department of the Interior Technical Guide.* Adaptive Management Working
Group, Washington, DC: U.S. Department of the Interior. Available at: www.usgs.
gov/sdc/doc/DOI-%20Adaptive%20ManagementTechGuide.pdf

54 Folke, C., et al. (2002). Resilience and Sustainable Development: Building Adaptive
Capacity in a World of Transformations. *AMBIO: A Journal of the Human Environment,* 31(5), pp. 437–440.

55 Gunderson, L. (1999). Resilience, Flexibility, Adaptive Management – Antidotes
for Spurious Certitude? *Conservation Ecology,* 3(1); McDonald, J. and Styles, M. C.
(2014) Legal Strategies for Adaptive Management Under Climate Change. *Journal of
Environmental Law,* 26(1), pp. 25–53.

56 Le Lièvre, C., et al. (2016) n. 50.

57 Holling, C. S., and Meffe, G. K. (1996) Command and Control and the Pathology of
Natural Resource Management. *Conservation Biology,* 10(2), pp. 328–337.

58 2014/52/EU (came into effect on 16 May 2017).

59 Article 8a(4).

60 Scottish Government/Marine Scotland (2012) *Survey, Deploy and Monitor Licensing
Policy Guidance (August 2012 version).* Edinburgh: Scottish Government. Available
at: www.scotland.gov.uk/Topics/marine/Licensing/marine/Applications/SDM. This is
currently the only such guidance, and Scotland is apparently the only jurisdiction
actively applying this approach (see O'Hagan, A. M. (2012) A Review of International Consenting Regimes for Marine Renewables: Are We Moving Towards Better
Practice? In: *Fourth International Conference on Ocean Energy.* Dublin: Ocean Energy
Systems.)

61 Scottish Government/Marine Scotland (2012) n. 60.

62 Bennet, F. (2016) *The Survey, Deploy & Monitor (SDM) Policy: An Example of Adaptive Management.* Webinar Annex IV, Tethys. Available at: www.youtube.com/
watch?v=VPC70eyWEtY

63 In fact, two cases: *R. v Rochdale MBC ex parte Milne* (No. 1) and *R. v Rochdale MBC
ex parte Tew* (1999) and *R. v Rochdale MBC ex parte Milne* (No. 2) (2000).

64 Infrastructure Planning Commission (2011) *Using the 'Rochdale Envelope' – Advice
Note 9.* Bristol: Infrastructure Planning Commission, p. 10. Available at: https://infra
structure.planninginspectorate.gov.uk/wp-content/uploads/2011/02/Advice-note-
9.-Rochdale-envelope-web.pdf

65 RenewableUK (2014) *Managing Regulatory and Consenting Costs for Offshore Wind.*
Report for RenewableUK by PMSS Consultants. London: RenewableUK, p. 83.

66 Fundingsland Tetlow, M., and Hanusch, M. (2012) Strategic Environmental Assessment: The State of the Art. *Impact Assessment and Project Appraisal,* 30(1), pp. 15–24.

67 Sustainable Energy Authority of Ireland (SEAI) (2010) *Strategic Environmental Assessment (SEA) of the Offshore Renewable Energy Development Plan (OREDP) in the Republic
of Ireland – Environmental Report.* Produced for SEAI by AECOM, Metoc and CMRC.
Available at: www.seai.ie/Renewables/Ocean-Energy/The-Offshore-Renewable-Energy-
Development-Plan-OREDP/Environmental_Report/

68 AECOM and Metoc (2009) *Offshore Wind and Marine Renewable Energy in Northern Ireland Strategic Environmental Assessment (SEA) Non-Technical Summary (NTS).*
Cheshire, England: AECOM Limited.

69 Department of Energy and Climate Change (DECC) (2011) *UK Offshore Energy
Strategic Environmental Assessment: OESEA2 Non-Technical Summary Future Leasing/*

Licensing for Offshore Renewable Energy, Offshore Oil & Gas, Hydrocarbon Gas and Carbon Dioxide Storage and Associated Infrastructure. London: DECC. Available at: www.gov.uk/government/publications/uk-offshore-energy-strategic-environmental-assessment-2-environmental-report

70 Ministerio de Industria, Energía y Turismo (2009) *Estudio Estratégico Ambiental del Litoral Español para la Instalación de Parques Eólicos Marinos.* Madrid: Ministerio de Industria, Energía y Turismo. Available at: www.aeeolica.org/uploads/documents/562-estudio-estrategico-ambiental-del-litoral-espanol-para-la-instalacion-de-parques-eolicos-marinos_mityc.pdf

71 Scottish Government (2008) *Post Adoption Statement: Scottish Marine Renewables SEA.* Edinburgh: Scottish Government, p. 232. Available at: www.gov.scot/Resource/Doc/921/0064201.pdf

72 Scottish Government (2007) *Scottish Marine Renewables Strategic Environmental Assessment Non-Technical Summary.* Report prepared for the Scottish Executive by Faber Maunsell and Metoc PLC. Available at: www.gov.scot/Resource/Doc/917/0048530.pdf

73 All documentation on Regional Location Guidance is available at: www.gov.scot/Topics/marine/marineenergy/Planning/windrlg

74 Scottish Government (2008) n. 71.

75 Scottish Government, Current Research Projects. Available at: www.scotland.gov.uk/Topics/marine/marineenergy/Research

76 Scottish Government, Scottish Marine Renewables Research Group (SMRRG). Available at: www.scotland.gov.uk/Topics/marine/marine-environment/smrrg

77 OEER (2008) *Fundy Tidal Energy Strategic Environmental Assessment Final Report.* Prepared by the OEER Association for the Nova Scotia Department of Energy. Halifax: NS Dept. of Energy, p. 92. Available at: www.oera.ca/wp-content/uploads/2013/06/FINAL-SEA-REPORT.pdf

78 Sinclair, A. J., Doelle, M., and Duinker, P. N. (2017) Looking Up, Down, and Sideways: Reconceiving Cumulative Effects Assessment as a Mindset. *Environmental Impact Assessment Review*, 62, pp. 183–194.

79 Doelle, M. (2009) The Role of Strategic Environmental Assessments (SEAs) in Energy Governance: A Case Study of Tidal Energy in Nova Scotia. *Journal of Energy & Natural Resources Law*, 27(2), pp. 111–144.

80 Department of Communications, Climate Action & Environment, Offshore Development Plan. Available at: www.dccae.gov.ie/en-ie/energy/topics/Renewable-Energy/electricity/offshore/offshore-renewable-energy-development-plan-/Pages/Offshore-Renewable-Energy-Development-Plan.aspx

81 92/43/EEC.

82 2009/147/EC.

83 Article 6(1), Habitats Directive.

84 Article 6(2), Habitats Directive.

85 Article 6(4).

86 European Commission (2010) *EU Guidance on Wind Energy Development in Accordance with the EU Nature Legislation.* Brussels: European Commission, p. 116. Available at: http://ec.europa.eu/environment/nature/natura2000/management/docs/Wind_farms.pdf

87 European Commission (2011) *Guidelines on the Implementation of the Birds and Habitats Directives in Estuaries and Coastal Zones with Particular Attention to Port Development and Dredging.* Brussels: European Commission, p. 46. Available at: https://ec.europa.eu/transport/sites/transport/files/modes/maritime/doc/guidance_doc.pdf

88 Ibid. p. 29.

89 O'Hagan, A. M. (2016). Marine Spatial Planning and Marine Renewable Energy. In: Copping, A., et al. (eds.) *Annex IV 2016 State of the Science Report: Environmental*

Effects of Marine Renewable Energy Development Around the World, Lisbon: Ocean Energy Systems, pp. 143–171. Available at: http://tethys.pnnl.gov/publications/state-of-the-science-2016

90 Therivel, R., and Ross, B. (2007) Cumulative Effects Assessment: Does Scale Matter? *Environmental Impact Assessment Review*, 27(5), pp. 365–385; Sinclair, A. J., Doelle, M., and Duinker, P. N. (2017) Looking Up, Down, and Sideways: Reconceiving Cumulative Effects Assessment as a Mindset. *Environmental Impact Assessment Review*, 62, pp. 183–194; Fundingsland Tetlow, M., and Hanusch, M. (2012) Strategic Environmental Assessment: The State of the Art. *Impact Assessment and Project Appraisal*, 30(1), pp. 15–24.

Part IV

Community and conflicts

Chapter 10

A sea of troubles? Evaluating user conflicts in the development of ocean energy

Jiska de Groot, Maria Campbell, Kieran Reilly, John Colton and Flaxen Conway

1 Introduction: ocean energy and marine user interactions – the potential for conflict

As an emerging industry, ocean energy (OE) is expected to play a major contributory role to reach long-term global emissions reductions targets. The offshore environment has become the new frontier for renewable energy production with many high-capacity wind parks and other technologies such as tidal and wave energy being increasingly developed. The scale and extent of the ocean renewables industry is unprecedented. Their demands for space places increased pressure on the marine area.[1]

Despite strong public support for renewable energy in principle, the terrestrial renewable energy sector has faced significant implementation barriers, in part driven by social acceptance and user conflicts.[2] Examples include conflicts between the wind energy sector and those using an area for tourism and recreation purposes,[3] large-scale solar PV developments displacing local livelihoods[4] and competition between the production of food and liquid biofuels.[5] These, among a number of other mostly technological, political and financial constraints,[6] have resulted in lengthy planning application procedures and high transaction costs for developers and the public sector. As a consequence of these, delivery of renewable energy targets has faced several setbacks.[7]

Expectations that the physical separation of OE from centres of population will remove issues of social acceptance and conflict have proven incorrect, as recent studies have demonstrated that moving renewable energy generation offshore is not an unproblematic alternative. Due to the potentially competing, objectives and priorities of OE and other industries, and in the delivering of conservation measures, the demand for space within our marine landscape is increasing. Interactions are therefore inevitable.[8] In addition to the significant technological and financial barriers the OE sector is facing, the interaction between the incoming sector and other users are important to consider in order to mitigate potential conflicts.[9] A key message from existing studies is that the offshore environment is not an empty space without other users, spatial claims or attachments.[10] Instead, it is an area often intensively used by local industries (including fishing, aquaculture transportation and tourism), and research or conservation zones have often been declared. Consequently, these spaces are often imbued with dependence, meaning and attachment. Commercial and recreational sectors, such as fishing, boating, or transportation, are important stakeholders operating in the marine environment and present a key sensitivity for OE deployment worldwide as they might face negative impacts from an incoming industry. Conflict between existing sectors in the marine space and the OE sector could stifle the development of this emergent sector.

This chapter focuses on the interactions and potential conflicts between the OE sector and other users of the sea, as the incumbent OE sector increases the demand for space within the marine landscape. To explore the tensions between the various uses of the marine space and identify potential areas for conflict, we will draw on case studies from North America, Ireland and the United Kingdom, exploring different areas of tension or potential conflict between the OE sector and existing uses. To conclude the chapter, we will discuss key themes emerging from the case studies and literature, together with some recommendations for mitigating conflict between the sectors.

2 User conflicts: which users of the marine space are likely to be affected by ocean energy development?

Despite the seeming endlessness of the marine area, the marine space in which the OE sector is likely to be situated is a busy place. Increasingly, spatial planning has extended from the terrestrial environment to include the marine environment.[11] Contributing to the potential for conflict is that the activities taking place in marine areas are often relatively near coastlines. As a consequence, a range of actors are likely to be using the marine space intensively and, for many, have already worked out how to coexist in shared space mostly by being able to use the same space at different times of the day or year or by using different parts of the space such as the surface/mid-water versus the bottom only.

To keep the momentum of building an offshore renewable energy sector, worldwide (large) areas of ocean space are being leased for development of offshore energy. Although with few exceptions most of these leases have been provided for the offshore wind sector, wave and tidal technologies have also been given leases in, among others, Ireland, Scotland, South Korea, France, the United Kingdom and North America. For example, in the United Kingdom since 2000, large areas of the UK seabed have been leased for offshore energy development with a generating capacity of up to 40 GW.[12] Although this was predominantly focused on offshore wind, in 2008 the first leasing round took place in Scotland for wave and tidal energy, which resulted in leases for six wave project development sites and four tidal stream sites to be leased with a potential of up to several 100 MW (megawatts).[13] This would restrict access to the leased marine areas for other users, possibly causing conflict.

Due to the high costs of operating in the marine space, it is likely that the OE sector will deploy technologies as close to population centres and near shorelines as possible to avoid costs and overcome technical constraints. Longer transmission distances lead to higher investment costs as well as higher to energy losses.[14] Yet this is also the area in which most other users operate, such as the local fishing and aquaculture sector, boating and recreation, research and conservation activities, and marine transport. Any incoming industry will therefore interact with an array of existing users of the ocean space, possibly causing user conflict or displacement. It is displacement that appears to be one of the most inflammatory conflicts between OE and existing uses, primarily due to the fact that OE use appears at this point to completely limit some other uses. Normally, exclusion zones are established around developments, which can significantly affect the distribution of fishing effort or may result in displacement altogether.[15]

We will now briefly discuss some of the key activities in the marine space that the OE sector is likely to interact with, with various consequences.

2.1 Commercial activity

2.1.1 Fisheries, aquaculture

Commercial fishers are a key group of ocean users who rely on access to fishing grounds for their livelihoods and are the group most likely to suffer direct adverse social and economic impacts as a result of spatial restrictions.[16] A case study into the potential barriers to the development of the Firth of Forth offshore wind farm noted that there would likely be a major challenge in overcoming opposition to the project from the fishing industry.[17] The idea that fishers are the main stakeholder group that may potentially be impacted by the development of OE has been claimed in a number of studies.[18]

The expansion of the OE sector may directly affect the fishing industry through loss of access to fishing grounds, reduced catchability of fish species during construction and operation, loss of gear or collisions with devices.[19] The potential negative impacts to fishers could lead to lead to conflict and opposition. At a more extreme scale, lack of acceptance could ultimately result in delays to projects through protests and protracted legal challenges. Fishers are a stakeholder group who have the ability to hinder projects through litigation, as was witnessed on the Cape Wind offshore wind farm. Cape Wind is a project to develop a 468-MW offshore wind energy farm off the coast of Cape Cod in Nantucket Sound, Massachusetts (USA). While not the only objection to this proposed project, the local fishers' organization objected on the grounds that it would restrict access to fishing grounds, make navigation more risky and prohibitively increase their costs.[20] The organization threatened the developer with legal action to have the wind farm moved. The federal lawsuit was dropped in 2012 after Cape Wind and the local fishers' association agreed to work together on ways to coexist as part of a settlement agreement.

The Cape Wind project ultimately failed for myriad other reasons, but this is an example of how addressing fishers' concerns (by ensuring that Horseshoe Shoal, the area proposed for the offshore wind farm, would remain open to fishing activities) is important. It shows that the acceptance or support of fishers is an essential component for the successful development of the OE sector. Despite these negative impacts and the potential for conflict,[21] OE projects could also provide a range of benefits and opportunities for fishers, such as alternative employment. Enhancing these benefits could help to increase the acceptance of projects. It should be noted that the acceptance of projects could lead to fishers sharing local knowledge, which would be beneficial to developers and could potentially contribute to the successful development of projects.

Fishers are major stakeholders in the marine environment with legitimacy, power and urgency in the decision-making process.[22] Despite this, commercial fishing is still considered the invisible sector.[23] It has been noted that the attitudes of the general public have tended to receive more attention in the academic literature, media and government publications,[24] and studies on the potential conflict between OE development and commercial fisheries are rare.

2.1.2 Other commercial activity

Other key commercial user groups of the marine space that could interact with the OE sector are the transportation, shipping, tug, fibre optic cables, and maritime service/safety sectors. It is important to recognize that much of the use of space on the ocean is 'temporary', while transiting through an area from point A to point B in the most efficient and safe manner. As previously stated, OE generation will be a more permanent occupation of the space. Nevertheless, observing the relationships, negotiations and agreements among existing ocean users and their efforts over the years to avoid or mitigate conflicts could inform the OE sector. For example, some ocean space – such as ferry routes and important shipping routes, especially entering and exiting harbours – would likely be deemed unfit for siting OE.[25]

Beyond proximity and place, like harbours, the other focus is the portion of the sea they are occupying during their temporary use. For example, ferries and shippers get from point A to point B occupying the surface and depth of draft for their vessels. Others, like fibre optic cables or commercial fishing, have this too, yet may also be focused the bottom of the sea. All will likely be resistant to adding additional fuel costs, communication, and possibly safety considerations needed in order to veer around an OE generation facility.[26]

To illustrate this point, tug-and-barge operators are similarly interested in the most direct route between points A and B, yet by their very nature of operation (e.g. towed barges can be about a quarter mile (1200 m) behind the tug), they've had to work cooperatively with other ocean users to avoid and mediate conflict. For example, in the waters off the Pacific coast of the United States since the 1970s, the coast's tribal and commercial crab fishers and tug-and-barge operators have recognized that they share the same sea lanes. These industries have worked together to avoid traffic disputes and costly collisions (vessels and gear) in year-round negotiations between crabbers and haulers along the 680-mile coast between San Francisco and Cape Flattery at Washington's northwest corner. This entirely voluntary process, undertaken without grant funding, has defined and continually refined pot-free shipping lanes and ship-free fishing areas, saving "many millions" along the way. Each year the agreements and lane maps are adjusted as needed. In Oregon in 2014–2015, for example, a new accord replaced two, 2-mile-wide tow lanes – an inshore lane open in the summer and fall and one offshore open in the winter and spring – with a single mile-wide, year-round lane located between the existing pair. The accord closed unused tow lanes into three minor ports – opening up new crabbing grounds – but reserved them in case tug traffic should resume. The key here was that these changes improved safety and reduced fuel costs for the tug operators, who saved miles and avoided stormier weather by moving closer to shore in winter, and the fishers lost access to the new single lane all year long but gained more than 175 square miles of new crabbing area in the old lanes. Both benefitted from greater certainty as to who belongs where. Similar agreements and an innovative working arrangement exist between the Oregon commercial fishing industry and the fibre optic cable industry.

One ocean user that is often overlooked but should be considered as an important stakeholder by the OE sector is marine science.[27] While these users are non-commercial in nature, they do have an attachment to the ocean as a place and, at times, over long temporal scales.[28] Conflicts would arise if special places or long-term studies could no longer be carried out due to the siting of an OE generation facility.

2.2 Recreational activity (boating, surfing, diving)

The OE sector has the potential to affect recreational activity. Thus far, studies investigating these interactions are few, but several areas of tension between the OE sector and recreational activity have already emerged.[29] The most visible of these tensions has been the potential conflict between surfing and harnessing wave energy, which has received significant attention both in the media and from several scholars.

Surfing is an important recreational and economic activity in many coastal areas, in which water users engage closely with diverse coastal environments. Crucially for the existence of surfing in an area is the existence of good-quality waves, which are created through complex interactions between shoreline structure, bottom conditions and available wave energy.[30] Small changes in local bathymetric conditions can change the quality of surf breaks, possibly rendering it impossible to surf.[31] It is therefore possible that the generation of OE may impact the surfing sector. Worldwide, many surfing areas are under threat due to restrictions in access, water quality problems, coastal development or disturbance to coastal ecosystems.

The surfing sector is expanding in intensity and the range of locations being used.[32] Several authors have argued that, despite its popularity, surfing is often seriously considered in coastal management decisions in part because, while this is slowly changing,[33] little is known about their local socio-economic impacts.[34] Because of the significant importance of the industry to many local areas, Lazarow et al.[35] call for clear articulation and measuring changes in recreational amenity and express the need to consider negative impacts on surf breaks that may result from development, coastal planning, protection works or other activities.

With regards to the OE sector, there is fear that the large-scale generation of wave energy may affect the quality of local surfing waves. This has prompted research into the potential effects on surfing waves resulting from wave energy generation and the perceptions of local surfers. These studies thus far found that among beach users, the perception was that the impacts would be minimal. Nevertheless, participants, in particular those depending on surfing for their incomes, were less positive when expressing their opinions and expressed concerns about changes in beach morphology and wave patterns.[36] This example will be further discussed as a case study in Section 3.2.

Similarly, the OE sector has the potential to affect other recreational activities in the marine environment,[37] including beach users. Fear of changes in coastal bathymetry and coastal erosion possibly resulting from OE deployment has been voiced. For example, changes in the bathymetry may alter wave and rip current patterns, affecting recreational beach users.[38] In addition, depending on the selected deployment site, wave and tidal may affect areas previously used by recreational boating and sailing, for example because of the exclusion zones surrounding the devices. Similarly, exclusion zones may affect the recreational scuba diving industry if large areas are being closed off.

3 Case studies

This section will discuss three international case studies in which the OE sector has or is perceived to cause conflict with existing users of the marine space. The first case study discusses fisheries conflicts in the island of Ireland, the second focuses on the surfing industry as an example of potential conflicts between tourism and leisure activities in the United Kingdom, and the third case study describes conflict between the OE sector and commercial fisheries.

3.1 Fisheries conflicts in the Island of Ireland

Three case studies were carried out on the attitudes and perceptions of fishers towards the development of OE on the island of Ireland (comprising Northern Ireland and the Republic of Ireland). The projects chosen were the Atlantic Marine Energy Test Site, the Torr Head and Fair Head tidal projects, and the First Flight Wind project (Figure 10.1). These sites represent offshore wind, wave and tidal energy.

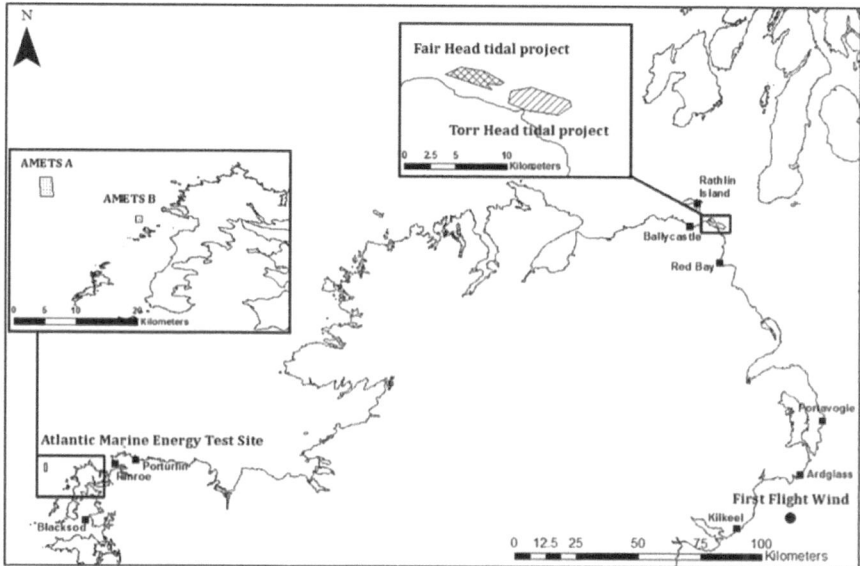

Figure 10.1 Fisheries conflicts case study sites

3.1.1 Atlantic Marine Energy Test Site

The Atlantic Marine Energy Test Site (AMETS) is being developed by the Sustainable Energy Authority of Ireland (SEAI) off the west coast of Annagh Head. The purpose of the site is to test and demonstrate the performance of wave energy devices in generating electricity and their survivability in open ocean conditions. The facility comprises two test sites: a near-shore test site (Test Site B), which will be utilised to test devices in 50-m water depth, and a site further offshore (Test Site A), which will test devices in 100-m water depth.[39]

Fishing activity in the Belmullet area consists mainly of crab and lobster fishers who operate small inshore vessels, most of which are between 10 m and 20 m.[40] Limited trawling is also carried out by up to three vessels that mainly operate from Killybegs (Donegal county) and are members of the Killybegs Fishermen's Organisation (KFO).[41] Two main fishing organizations operate in the area, the largest of which is the Erris Inshore Fishermen's Association (EIFA), counting roughly 65 registered members, whilst the Erris Lobster and Crab Re-stocking Association (ELCRA) counts eight registered fishing vessels.

3.1.2 Fair Head/Torr Head tidal projects

In October 2012, Tidal Ventures was awarded two leasing agreements that give them exclusive rights to investigate the development of a commercial-scale

100-MW tidal energy array in the waters offshore of Torr Head on the north coast of Northern Ireland, as well as the development of a 100-MW tidal energy project at Fair Head. It is estimated the project will occupy an area of approximately 3 km². Fishing activity in the vicinity of the proposed projects is similar. Members of the North Coast Lobster Fishermen's Association (NCLFA) fish regularly in both areas for lobster and crab. As the project sites are located in relatively close proximity to each other, it was assumed that the same group of fishers could potentially be impacted by both. As such, the Torr Head and Fair Head tidal projects were treated as a single case study site.

3.1.3 First Flight Wind

In October 2012, the First Flight Wind (FFW) consortium[42] was selected to develop a 600-MW offshore wind farm in an allocated wind resource zone in Northern Ireland. In June 2014, the target capacity of the project was reduced to between 300 MW and 400 MW due to the significant extent of fishing and navigation interests in the area. In December 2014, it was announced that the project was being cancelled due to the fact that it would not be completed in time to benefit of the UK renewable energy market incentives, which stated that construction of the project would have had to be completed by 2020.

The majority of the fishing fleet in Northern Ireland is based at three ports along the County Down coast – Kilkeel, Ardglass and Portavogie.[43] There are also other smaller ports along the County Down coast where smaller numbers of fishers operate from. This includes Annalong and Newcastle. Commercial fishing in the area around the proposed project is diverse, and a variety of fishing methods, such as trawling and crab/lobster potting, are used to target a wide range of species. The most important catch for the majority of vessels is *Nephrops norvegicus*, or Dublin Bay prawns, which are landed on a daily basis into local ports.

3.1.4 Anticipating user conflict: fishers' attitudes towards ocean energy

Surveys were carried out in each of the preceding sites described exploring the attitudes and perceptions of fishers towards the development of ocean energy locally. The large majority (93%) of survey respondents indicated that they were aware of the proposed OE development near their homeports. Responses to the statement 'I think that it is important to develop OE resources such as offshore wind, wave and tidal energy' were divided with 40% either strongly agreeing or agreeing with the statement, compared to 45% who disagreed. A further 15% of the respondents were neutral on the statement (Figure 10.2).

Figure 10.2 further shows a notable difference in fishers' attitudes towards OE (explained as marine renewable energy (MRE) in the survey) at site level. The AMETS site was considered the most acceptable site for OE with 55%

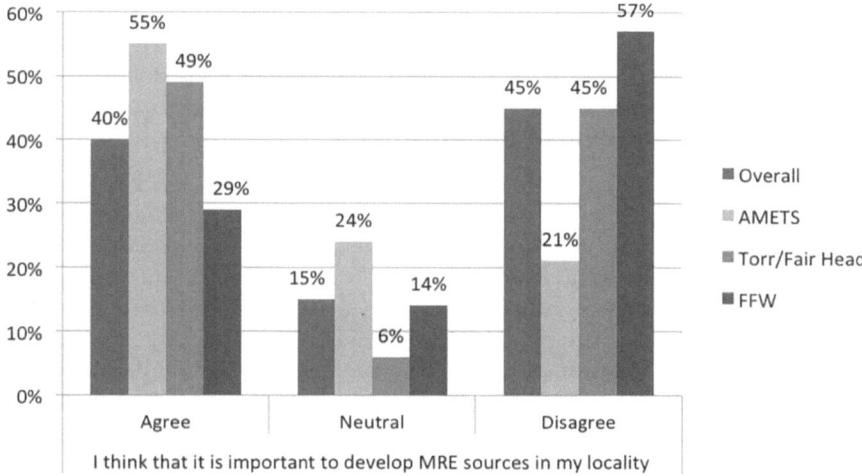

Figure 10.2 Attitudes of fishers towards ocean energy

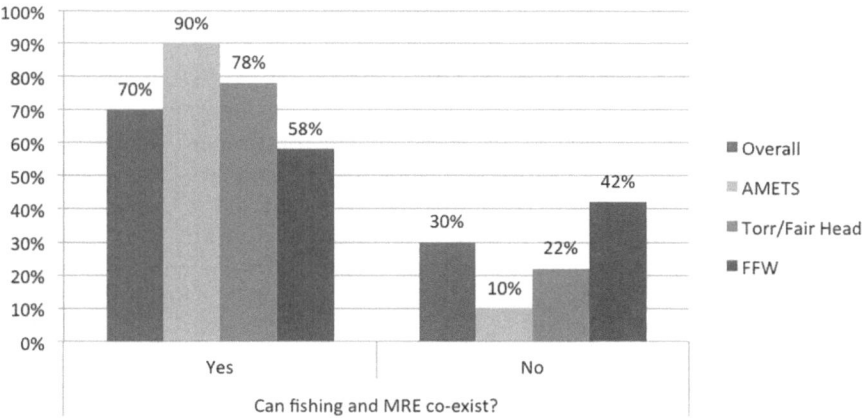

Figure 10.3 Coexistence of the ocean energy sector and commercial fishing

of those surveyed in agreement that it is important to develop OE, while only 29% of the sample at the FFW site agreed with this same statement. Similarly, when asked if it would be possible for the fishing and OE industries to coexist, the majority of respondents (70%) were of the opinion that this could be possible (Figure 10.3).

The coexistence responses follow a similar pattern to the question about attitudes, in that the highest level of agreement was at AMETS and the lowest level at the FFW site. These answers may be driven by both project-related factors[44] and local contextual factors,[45] such as the type of fishing in the area and gear type.[46] Therefore, the potential for conflict and coexistence may differ between sites.

3.2 Anticipating user conflicts in the United Kingdom: the case of surfing and the wave hub

The Wave Hub facility in Cornwall (South West UK) is a 48-MW OE test site, predominantly designed for the purpose of trialling wave energy converters prior to commercialisation. Developers can test devices in open sea conditions, and the four testing berths allow for testing of both single devices and arrays. Furthermore, the site is 16 km offshore from the nearest town (Hayle), is grid connected and fully consented with a 25-year seabed lease.[47] Cornwall is a peripheral rural location in the South West of the UK, which, despite various development programmes, has experienced almost continual socio-economic decline for more than a century.[48] The area is largely dependent upon agriculture, tourism and fisheries. Recreational water users, including bathers and surfers, are attracted by Cornwall's pristine beaches. Beach water users such as bathers and surfers are of high economic importance to Cornwall. A study by leading not-for-profit organization Surfers Against Sewage (SAS) estimated the total spending of surfers in the area at over £100 million per year[49] and that it was responsible for over 1,500 jobs.[50]

Because of the 'invisibility' of the surfing sector (e.g. due to times of beach visits – early and late in the evenings – in which general tourist surveys are rarely conducted,[51] the surfing community is concerned that its interests and views are being ignore or overlooked. This is illustrated by the strategic environmental assessment (SEA) process for offshore energy concerning a seabed leasing plan for offshore energy development. The report published[52] has no reference to either surfing or recreation more broadly, and its annex covering the environmental baseline for tourism and recreation contains a few anecdotal references to surfing in some areas rather than a serious consideration of the locally highly significant industry.[53]

The surfing community in the UK has expressed concerns that offshore renewable energy threatens surfing resources and recreation. With regard to the OE sector, the concerns have promoted research into the potential for conflict between the surfing sector and wave energy generation through an investigation into potential changes to wave height and period, as well as wave characteristics preferred by the surfing sector.[54]

Some of the most important anticipated impacts on surfing and recreation will happen if offshore renewable energy developments affect coastal wave dynamics and coastal morphology. For example, the technology could weaken incoming waves, change the direction of the incoming wave, cause interference between waves or cause differences in the breaking of the waves at the beach.[55] A suitable wave climate is essential for both the existence and quality surfing waves and is therefore indispensable for the surfing industry. The incoming OE sector may affect the wave climate. Structures can cause wave energy to be blocked, altered or extracted as it passes through a development site. Impacts on wave resources may occur when the technology interrupts the waves' shoreward transmission and changes their characteristics that these changes are manifest in the surf zones when the waves break there.[56] Impacts have already been felt from offshore wind farm development. For example, the London Array reported that changes to the wave climate occur as turbine structures scatter the incoming wave. These are all factors that, if they occur, could cause damage to the surfing and recreation sectors.[57] Other studies that have focused on these aspects of wave energy indicate that any loss of wave energy transmitted (e.g. as a result of a wave energy generator) typically will cause a reduction in wave height and, to a lesser extent, a change to wave period.[58]

Although studies to date suggest that the impacts of offshore wind farm developments and tidal stream demonstrator projects on surfing may be negligible and that the impacts of wave demonstrator projects are small, surfers' concerns are likely to remain until proven otherwise by monitoring.[59] A key concern from the surfing sector is that despite 'negligible' impacts on the surfing sector at the current level of OE activity, little is known about the potential impacts of the future developments proposed to deliver long-term renewable energy targets. Because those will be substantially larger than those studied to date, the sector fears that the impacts are also likely to increase.[60]

During the development of the Wave Hub, several concerns were raised with regard to the potential socio-economic impacts of the test facility on both tourism and fisheries sectors. For example, during the Wave Hub consultation, concerns were raised among stakeholders that wave energy extraction would reduce the height and quality of coastal waves for surfing, as well as affecting sediment transport and beach morphology.[61] In 2007, however, Millar et al.[62] predicted impacts of the Wave Hub and claimed that average wave height reductions would be 1 cm and the maximum wave height reduction would be up to 3 cm for the popular surfing locations along the north of coast of Cornwall. This is an important finding because, during the consultations for the Wave Hub, serious concern was expressed by the surfing sector that the development would seriously impact local surfing areas in particular through a reduction in wave height. In the same year that Millar et al.'s[63] paper was published, the developers of the Wave Hub were presented with a petition signed by 500 surfers who raised concerns that the wave energy converters proposed to be trialled in the Wave Hub might affect water recreation and tourism on the beaches in its lee.[64] The primary concern of this user group was the possibility of reduced inshore wave height and effects of wave periods.

To explore the specific concerns of these user groups, two studies by Stokes et al.[65] explored preferred surf conditions by various beach water user groups and explored how surfers and other beach users perceived physical coastal impacts caused by the Wave Hub. A key finding from the latter investigation was that

surfers (and expert water users) were among the groups most likely to anticipate negative impacts to recreational wave amenity, with a key reason that they
already perceive ideal wave conditions for recreation to be scarce. As one of
Stokes et al.'s participants put it: "Surfing is a fickle thing, you only get those few
days a year where it's that good, so you want to keep that, you know maximise
that as best you can".[66] His study found that wave energy deployments that are
perceived to be large scale, close to shore, wide, stationary or extracting high
percentages of energy are likely to invoke anticipations of significant or severe
coastal impacts. Conversely, those that are perceived to be small scale, far from
shore, narrow, moving or extracting low percentages of wave energy are more
likely to invoke anticipations of insignificant or no coastal impact.[67] Crucial for
causing potential conflict between the OE sector and surfing was the anticipation
among participants that OE could cause a range of impacts on various elements
of the coastal environment. Although some participants expected none at all,
the main impacts discussed were reduced wave height or wave quality, changes
in sediment transport, coastal erosion, changes in rip current behaviour, and the
possibility of a device breaking free and washing ashore.[68]

More concerning and potentially conflict causing are the findings of a similar study that the preferred wave period of all water users in the sample was
approximately equal to the peak period associated with the bulk of available wave
energy.[69] Because technologies are likely to extract energy efficiently only in a
finite range of frequencies, there might be a clash of interest between device
developers and water users.[70] Consequently, the authors further indicate that "if
water users learn that energy extraction will target waves of around 12 s, there
may be enhanced opposition from this stakeholder group".[71]

To avoid conflict between the two sectors, it is therefore important to clearly
communicate the low levels of likely height reductions of energy extraction
(which is predicted to be <0.5% in a scenario of 30% energy extraction.[72]
Engagement between the two sectors is ongoing. Key stakeholders in the surfing
sector that have started to engage with the OE sector are Surfer's Against Sewage (SAS), a not-for-profit organization campaigning for clean, safe recreational
waters; the British Surfing Association (BSA), the governing body for surfing
in the UK; and the British Professional Surfing Association (BPSA), which
provides a platform for surfers and sponsors to raise their problems in the surfing
industry and showcases the quality of surf conditions and locations across the
country.[73]

An important position of SAS is that they need to become stakeholders in the
OE debate (and offshore energy more broadly). They argue that surfers and other
water users need an official recognition in the policy arena surrounding relevant
issues, such as marine management, offshore energy siting and conservation.[74]
The exemplary case referred to is the acknowledgement from the Scottish Government that recreational water users need a voice in decisions concerning the
marine environment. As a result of extensive lobbying by the SAS, the Scottish
Government granted these users a seat on regional planning partnerships under

the Scottish Marine and Coastal Access Act.[75] This has created a platform in which surfers and other beach users can engage one another and voice concerns. Further political efforts are made by SAS that set up the Protect Our Waves petition to achieve better political engagement on the matter.[76] Such engagement could contribute to mitigate potential conflicts between the surfing and recreational sector and the OE sector.

3.3 Tidal energy development in Nova Scotia: conflict with commercial fishery

The Bay of Fundy in Nova Scotia, Canada, has the highest tidal range in the world. Given this extensive range, tidal energy development has been explored for almost a century. Yet only recently have there been significant efforts to harness this energy. In the mid-1980s the Annapolis River Tidal Generating Station was constructed in barrage fashion with a single turbine. Impacts of this development on fish have been significant and include fish mortalities and shortened lifespan.[77] Experience with this project, including lack of meaningful and effective mitigation strategies, has impacted, in part, the experience and attitudes of fishers with more recent tidal energy development in the Bay of Fundy.

The New Minas Passage, a constriction in the upper Bay of Fundy, is the site of the Fundy Ocean Research Centre for Energy (FORCE). Established in 2009, FORCE serves as the hub for in-stream tidal energy turbine testing hosting four berths.[78] In 2009, the first Open Hydro test turbine was deployed in the Minas Passage. Key stakeholders in the region, including commercial fishing groups, aboriginal communities, local municipalities, and other stakeholders in the region, did not contest this deployment at the time. The turbine was removed relatively shortly after deployment when it was discovered that it had received significant damage.[79] In the lead-up to their second deployment of a 2-MW turbine planned for spring 2016, Cape Sharp (now a joint venture between Emera and OpenHydro) encountered significant pushback from a key stakeholder group, the commercial fishery represented by the Bay of Fundy Inshore Fishermen's Association (BIFA).

Cape Sharp plans for harnessing energy from the New Minas Passage are significant. Following the initial deployment of 4 MW (two 2-MW turbines) and providing there is regulatory approval, there are plans for commercial scale-up. This will include 16 MW in 2017, 50 MW (17 more turbines) in 2019, growing up to 300 MW (150 turbines) by the mid-2020s. This is enough generating capacity for 75,000 households. BIFA has concerns regarding this development and the movement from the FORCE site as a test centre to a centre supporting commercial scale-up. Key concerns among BIFA are that the turbines will cause environmental damage including fish mortalities and that the monitoring programme is insufficient. Their position was supported, in part, by the report from the Canadian Department of Fisheries and Oceans (DFO), *Review of the Environmental Effects Monitoring Program for the Fundy Tidal Energy Project*.[80] In this report, the

DFO notes that FORCE's monitoring plan outlined in the 2015 environmental effects monitoring program (EEMP) was lacking in the following areas:

- Inability to monitor mortality
- Lack of knowledge of turbine interactions with fish
- Gaps in baseline data
- Lack of accounting for spatial and temporal changes
- Lack of accounting for effects of scaling up
- Lack of technology specifications
- Lack of mitigation measures

Since this report was released, FORCE, with financial support from Government, has developed an effective environmental monitoring programme to address these issues.

The Nova Scotia Department of Environment announced in June 2016 that Cape Sharp Tidal was approved to deploy their turbines, but the company continued to delay deployment, citing ongoing consultation with BIFA and other commercial fishers.[81] Colin Sproul, the spokesman for the Bay of Fundy Inshore Fisherman's Association (BIFA), noted in an interview[82] that, while his association would be interested in engaging with Cape Sharp Tidal, the company's engagement style did not sit well with him: "[W]e just feel it's irresponsible to do it in the off-the-record, informal manner, like they've requested".

In July 2016, the Bay of Fundy Inshore Fisherman's Association, representing 175 fishers, applied to the Nova Scotian Supreme Court to repeal the province's decision to approve the deployment of two tidal turbines in the Bay of Fundy.[83] The key concerns of BIFA were the lack of monitoring, lack of baseline research and concern over fish and marine mammal interactions with the turbines.[84] The motion by the Bay of Fundy Fisherman's Association was denied by the Nova Scotia court in late October 2016. Another motion to delay the deployment of two 2-MW OpenHydro turbines went before the Nova Scotia Supreme Court in February 2017. The focus of this motion was the monitoring programme established by FORCE. The court has yet to announce its decision.

Given that these events have occurred relatively recently, literature related to these events is scarce other than what has been recorded in local media. Insight for this brief case study is, in part, drawn from first-hand accounts of public meetings, participation in government and industry stakeholder meetings, and involvement as co-chair of the Fundy Energy Research Network (FERN), a group of stakeholders from industry, academia, government (both municipal and provincial) and NGOs.

Leading up to these events, representatives of the fishing community expressed frustration over the lack of consultation by FORCE. While members of the fishing community were consulted during the strategic environmental assessment developed in 2008, ongoing consultation after the SEA was limited. This was due, in part, to lack of effort by FORCE but also to the unorganized nature of the commercial fishing community. Key principles and processes highlighted in the SEA were the precautionary principle and adaptive management. Fishers represented by the Bay of Fundy Inshore Fishermen's Association (BIFA) have been quick to point out that these principles do not appear to have guided the more recent events related to tidal energy development. Plans for commercial scale-up to 300 MW is one example noted by BIFA where they were not consulted.

The ensuing conflict has garnered significant media attention including opinion pieces for and against, written editorials and even satirical and/or editorial cartoons. In some cases, what is reported is misinformation related to the impacts of tidal turbines on fish and marine mammals, causing broader public negative attitudes towards tidal energy development in Nova Scotia. For example, the spokesperson for BIFA was quoted in a provincial paper as saying up to 3,500 turbines would be deployed in the Minas Passage. This was clearly a misleading statement, especially when the Nova Scotia Government set a potential maximum target of 300 MW, which could be achieved with 150 turbines. While direct contact with turbines could result in harm to fish, studies in the East River (New York), Cobscook Bay (Maine) and European Marine Energy Centre (Scotland) have indicated that fish have not made contact with an operating turbine.[85] In fact, in one study, fish avoidance behaviour was noted 140 m upstream of an operating turbine.[86]

The end result to date is a lack of trust by BIFA and the members it represents towards the Fundy Ocean Research Centre for Energy (FORCE), scholars engaged in tidal energy research and government. Given this state, it is important to consider how to reset, if this is possible, the relationships among the key players in this conflict and to renew constructive dialogue in order to advance tidal energy development in a way that rebuilds trust and respect. A first step in this process would be to recognize more deeply the connections to *place*,[87] developed by Bay of Fundy fishers, and to what extent this connection and the knowledge derived from it can offer guidance to policy development related to tidal energy development. Critically examining the role of *power* and how it is distributed, wielded and shared would support a renewed consultation process as well.

4 Discussion and concluding remarks: challenges and opportunities for coexistence

The literature review and case studies have revealed a range of possible conflicts between the OE sector and sectors that already operate in the marine space. The chapter has shown that, although people generally like the idea of renewable energy, challenges arise when the question is where to site the devices and at what scale they are to be deployed. Whereas the OE is often presented as a 'frontier' of energy generation that is 'out of sight and out of mind', this is simply untrue. Similarly, perceptions of the ocean as a frontier are a misnomer: the marine space is a busy place, and many people depend on the space for their livelihoods. As a consequence, the location, size and potential impacts on existing uses that new developments may have become important for mitigating conflict. This section discusses the key themes that emerged from the case studies and literature study.

Firstly, local and place contextual factors are important for understanding or mitigating conflict. Different users of the marine space use the space differently and are therefore differently affected by an incoming OE sector. For example, surfing operates very near shore, tugs and dredges in the near shore, and fishers operate both near shore and further out. Similarly, some users operate on the surface, whereas others operate under water. Understanding where different stakeholders are operating in relation to the OE site is important to avoid conflict. Furthermore, such understanding will provide insight into who needs to be engaged in the siting process.

Secondly, insight into the interaction and displacement of existing uses is crucial to understand potential for conflict. This requires insight into whether the space can be shared or if restricted or exempt use is required for a particular activity. Depending on the type of device or local regulatory framework, the incoming OE sector may cause complete or partial displacement of sectors or activities, thereby significantly impacting other users. In the context of renewable energy siting, displacement is generally caused by exclusion zones. Discussed mainly in the context of fisheries, it can be a serious cause of conflict.

In some cases, it was expected that the two sectors could exist peacefully if potential negative impacts are managed. For example, tug and towing operations, as well as leisure activities near shore, could share the ocean space with other users. Similarly, fisheries and OE sector could coexist if through appropriate stakeholder engagement decisions are made that safeguard existing fishing areas from exclusion zones or allow for some fishing in the area.[88] However, despite these promising findings, the potential for conflict between fisheries and the OE sector should not be underestimated, as displacement is likely to increase with the size of developments. In addition, it should be recognised that coexistence may lead to additional costs for one or more users to operate under a business-as-usual scenario. Examples of this are additional fuel costs for ferries and ships if they need to travel around large exclusion zones. To account for existing users of the sea and to avoid displacement, the OE sector needs to engage with key affected stakeholders and take existing uses into account when determining the optimal siting of devices.

Thirdly, a potential cause of significant conflict would be if the OE sector alters the physical characteristics of its site, thereby altering the local resource that existing industries are dependent on. The example of the surfing industry in Cornwall illustrated this potential cause for conflict. Although at present the perceived impacts on the local surfing and beach resource are manageable, *if* the OE sector would affect the local bathymetry or beach characteristics –, including rip currents, reduced wave action and/or coastal erosion – this could cause significant conflict. Such conflict would be much more difficult to manage as the resultant impacts are of a more permanent nature. In addition, the danger of the 'reduced quality of experience', as might be experienced by the surfers, was another cause of concern that needs to be carefully managed in the case of larger arrays of, in particular, wave energy technologies. Therefore, these risks need to be carefully investigated and managed before placing devices in areas with high dependence on a particular industry.

Fourthly, a key area for understanding and mitigating conflict concerns economic factors. This requires insight into who benefits economically from a siting and who might be affected. For example, an OE development may create jobs for one industry at the cost of job losses for another. The example of the fishing industry illustrated that, although job losses may happen, there is the potential transferability of jobs into the OE sector. Whereas this may not be the case in tug and towing operations, they might derive some benefit from support for and/ or enhancement of local infrastructure. The case of the surfing and recreational sectors is more complicated as a potentially reduced quality of experience in recreational activities may impact on businesses operating in the sector, as well as the 'recreational end user'. Although the latter is unlikely to face direct economic impacts, the enjoyment they derive from the resource will be compromised in a way that is not possible to express in monetary terms.

Finally, governance, policy, and planning are key for understanding and avoiding conflict particularly in relation to the power to influence decision-making processes. An important aspect of this concerns the link between the location and siting of devices and the power of the affected users to influence decisions. For example, it is highly unlikely that devices will be placed near the opening of a major port or shipping lane, yet they could be placed in the middle of a prime crab port. This raises questions about who has a voice in decision making and what powers are granted to each affected sector. To illustrate, the surfing sector may be among the users affected by OE siting; yet, they have a limited voice in siting processes as a marine user. Therefore, who has the power to affect siting and who does not have the power become important questions in trying to understand siting conflict. Other tensions also exist that need to be governed, including the tension between a return on investment and the expense of long legal battles for developers and authorities, and affordability of energy for end users. A final issue that should not be forgotten is that policy, governance and planning themselves are prone to internal conflict. For example, because of a range of national and international commitments, the precautionary approach to energy siting as part of an impact assessment needs to be balanced against climate change pressures. Although these are difficult to influence, thorough and genuine engagement during decision-making processes may reduce conflict or avoid escalation.

In conclusion, this chapter has demonstrated that the incoming OE sector needs to take into account a range of existing users of the marine space. Insight into local and context-specific factors that cause conflict, the potential for coexistence, the nature of the potential impacts and who has a voice in siting decisions were identified as being important for mitigating conflict among the sectors.

Failing to do so could result in serious conflict among existing users of the marine space and the emerging sector, thereby stifling its development.

Notes

1 Wright, G. (2015) Marine Governance in an Industrialised Ocean: A Case Study of the Emerging Marine Renewable Energy Industry. *Marine Policy*, 52, pp. 77–84.
2 Bell, D., Gray, T., and Haggett, C. (2005) The 'Social Gap' in Wind Farm Siting Decisions: Explanations and Policy Responses. *Environmental Politics*, 14(4), pp. 460–477.
3 Lilley, M. B., Firestone, J., and Kempton, W. (2010) The Effect of Wind Power Installations on Coastal Tourism. *Energies*, 3(1), p. 1.
4 Yenneti, K., Day, R., and Golubchikov, O. (2016) Spatial Justice and the Land Politics of Renewables: Dispossessing Vulnerable Communities Through Solar Energy Megaprojects. *Geoforum*, 76, pp. 90–99.
5 Rathmann, R., Szklo, A., and Schaeffer, R. (2010) Land Use Competition for Production of Food and Liquid Biofuels: An Analysis of the Arguments in the Current Debate. *Renewable Energy*, 35(1), pp. 14–22.
6 Magagna, D., and Uihlein, A. (2015) Ocean Energy Development in Europe: Current Status and Future Perspectives. *International Journal of Marine Energy*, 11, pp. 84–104; Rourke, F. O., Boyle, F., and Reynolds, A. (2010) Marine Current Energy Devices: Current Status and Possible Future Applications in Ireland. *Renewable and Sustainable Energy Reviews*, 14(3), pp. 1026–1036; Bahaj, A. S. (2011) Generating Electricity from the Oceans. *Renewable and Sustainable Energy Reviews*, 15(7), pp. 3399–3416.
7 Cotton, M., and Devine-Wright, P. (2011) NIMBYism and Community Consultation in Electricity Transmission Network Planning. In: Devine-Wright, P. (ed.) *Renewable Energy and the Public*. London: Earthscan, pp. 115–128.
8 Wright, G. (2015) n. 1.
9 Campbell, M. S. (2017) *Fisheries, Marine Conservation, Marine Renewable Energy and Displacement: A Fresh Approach*. Plymouth, UK: University of Plymouth.
10 Bailey, I., West, J., and Whitehead, I. (2011) Out of Sight but Not out of Mind? Public Perceptions of Wave Energy. *Journal of Environmental Policy & Planning*, 13(2), pp. 139–157; Conway, F., et al. (2010) Ocean Space, Ocean Place: The Human Dimensions of Wave Energy in Oregon. *Oceanography*, 23(2), pp. 82–91; Devine-Wright, P. (2009) Fencing in the Bay? Place Attachment, Social Representation of Energy Technologies and the Protection of Restorative Environments. In: Bonaiuto, M., et al. (eds.) *Urban Diversities, Biosphere and Well-being: Designing and Managing Our Common Environment*. Hogrefe & Huber.
11 Stelzenmüller, V., et al. (2013) Monitoring and Evaluation of Spatially Managed Areas: A Generic Framework for Implementation of Ecosystem-Based Marine Management and Its Application. *Marine Policy*, 37, pp. 149–164; Douvere, F., and Ehler, C. N. (2009) New Perspectives on Sea Use Management: Initial Findings from the European Experience with Marine Spatial Planning. *Journal of Environmental Management*, 90, pp. 77–88; Halpern, B. S., et al. (2012) Near-term Priorities for the Science, Policy and Practice of Coastal and Marine Spatial Planning. *Marine Policy*, 36, pp. 198–205.
12 The Crown Estate (2013) *Leasing Rounds and Projects*. Available at: www.thecrown estate.co.uk/energy-infrastructure/wave-and-tidal/pentland-firth-and-orkney-waters/ leasing-round-and-projects/
13 The Crown Estate, *Wave and Tidal*. Available at: www.thecrownestate.co.uk/ energy-minerals-and-infrastructure/wave-and-tidal/

14 Gasch, R., and Twele, J. (2001) *Wind Power Plants: Fundamentals, Design and Operation*. Germany: Solar Praxis AG.

15 Alexander, K. A., Potts, T., and Wilding, T. A. (2013) Marine Renewable Energy and Scottish West Coast Fishers: Exploring Impacts, Opportunities and Potential Mitigation. *Ocean & Coastal Management*, 75, pp. 1–10; Mackinson, S., et al. (2006) A Report on the Perceptions of the Fishing Industry into the Potential Socio-Economic Impacts of Offshore Wind Energy Developments on Their Work Patterns and Income. In: *Science Series Technical Report*. Lowestoft: CEFAS; Hiddink, J. G., et al. (2006) Predicting the Effects of Area Closures and Fishing Effort Restrictions on the Production, Biomass, and Species Richness of Benthic Invertebrate Communities. *ICES Journal of Marine Science: Journal du Conseil*, 63(5), pp. 822–830; de Groot, J., et al. (2014) Investigating the Co-existence of Fisheries and Offshore Renewable Energy in the UK: Identification of a Mitigation Agenda for Fishing Effort Displacement. *Ocean & Coastal Management*, 102, Part A, pp. 7–18; Pomeroy, C., Hall-Arber, M., and Conway, F. (2015) Power and Perspective: Fisheries and the Ocean Commons Beset by Demands of Development. *Marine Policy*, 61, pp. 339–346; Sullivan, C. M., et al. (2015) Combining Geographic Information Systems and Ethnography to Better Understand and Plan Ocean Space Use. *Applied Geography*, 59, pp. 77–79.

16 Yates, K. L., and Schoeman, D. S. (2013) Spatial Access Priority Mapping (SAPM) with Fishers: A Quantitative GIS Method for Participatory Planning. *PLoS ONE*, 8(7), p. e68424.

17 O'Keeffe, A., and Haggett, C. (2012) An Investigation into the Potential Barriers Facing the Development of Offshore Wind Energy in Scotland: Case Study – Firth of Forth Offshore Wind Farm. *Renewable and Sustainable Energy Reviews*, 16(6), pp. 3711–3721.

18 Alexander, K. A., Potts, T., and Wilding, T. A. (2013) n. 15; Alexander, K. A., Wilding, T. A., and Heymans, J. J. (2013) Attitudes of Scottish Fishers Towards Marine Renewable Energy. *Marine Policy*, 37, pp. 239–244; Gray, T., Haggett, C., and Bell, D. (2005) Offshore Wind Farms and Commercial Fisheries in the UK: A Study in Stakeholder Consultation. *Ethics, Place & Environment*, 8(2), pp. 127–140; Rodmell, D. P., and Johnson, M. L. (2002) The Development of Marine Based Wind Energy Generation and Inshore Fisheries in UK Waters: Are They Compatible? In: Hart, P., and Johnson, M. (eds.) *Who Owns the Sea? Workshop Proceedings*. Scarborough: University of Hull, pp. 76–105; Fayram, A. H., and de Risi, A. (2007) The Potential Compatibility of Offshore Wind Power and Fisheries: An Example Using Bluefin Tuna in the Adriatic Sea. *Ocean & Coastal Management*, 50(8), pp. 597–605; Hall, D. M., and Lazarus, E. D. (2015) Deep Waters: Lessons from Community Meetings About Offshore Wind Resource Development in the U.S. *Marine Policy*, 57, pp. 9–17.

19 Campbell, M. S. (2017) n. 9; Perry, K., and Smith, S. L. (2012) *Rhode Island Ocean Special Area Management Plan: Fisheries Mitigation Options – A Review*. Kingston: Coastal Resources Centre, University of Rhode Island.

20 Seccombe, M. (2010) *Fearing Drastic Losses, Fishermen Dodge Legal Traps in Cape Wind Fight*. Available at: http://mvgazette.com/news/2010/12/09/fearing-drastic-losses-fishermen-dodge-legal-traps-cape-wind-fight?k=vg5343e13f8dc57

21 Alexander, K. A., Potts, T., and Wilding, T. A. (2013) n. 15.

22 de Groot, J., et al. (2014) n. 45; Pomeroy, C., Hall-Arber, M., and Conway, F. (2015) n. 15; Johnson, T., Jansujwicz, J., and Zydlewski, G. (2015) Tidal Power Development in Maine: Stakeholder Identification and Perceptions of Engagement. *Estuaries and Coasts*, 38(1), pp. 266–278.

23 Pomeroy, C., Hall-Arber, M., and Conway, F. (2015) n. 15.

24 O'Keeffe, A., and Haggett, C. (2012) n. 17.

25 Industrial Economics (2012) *Identification of OCS Renewable Energy Space-Use Conflicts and Analysis of Potential Mitigation Measures*. Herndon, VA: Department of the Interior, Bureau of Ocean Energy Management.

26 Ibid.

27 Sherman, K., et al. (2016) The Oregon Nearshore Research Inventory Project: The Importance of Science and the Scientific Community as Stakeholders in Marine Spatial Planning. *Ocean & Coastal Management*, 130, pp. 290–298.

28 Industrial Economics (2012) n. 25; Sherman, K., et al. (2016) n. 27.

29 SAS (2009) *Guidance on Environmental Impact Assessment of Offshore Renewable Energy Development on Surfing Resources and Recreation*. St Agnes, Cornwall: Surfers Against Sewage.

30 Scarfe, B. E., et al. (2003) The Science of Surfing Breaks – A Review. In: *Proceedings of the 3rd International Surfing Reef Symposium (June 22–25)*. Hamilton: ASR Ltd.

31 SAS (2009) n. 29; Nelsen, C., Cummins, A., and Tagholm, H. (2013) Paradise Lost: Threatened Waves and the Need for Global Surf Protection. *Journal of Coastal Research*, 65, pp. 904–908.

32 Lazarow, N., Miller, M. L., and Blackwell, B. (2008) The Value of Recreational Surfing to Society. *Tourism in Marine Environments*, 5(2–1), pp. 145–158.

33 Martin, S. A., and Assenov, I. (2014) Investigating the Importance of Surf Resource Sustainability Indicators: Stakeholder Perspectives for Surf Tourism Planning and Development. *Tourism Planning & Development*, 11(2), pp. 127–148.

34 Nelsen, C., Cummins, A., and Tagholm, H. (2013) n. 31; Lazarow, N., Miller, M. L., and Blackwell, B. (2008) n. 32; Eardley, C. S., and Conway, F. (2011) *Oregon's Non-consumptive Recreational Ocean User Community – Understanding an Ocean Stakeholder*. Corvallis, OR: Oregon Sea Grant Publication.

35 Lazarow, N., Miller, M. L., and Blackwell, B. (2008) n. 32.

36 Stokes, C., et al. (2014a) Anticipated Coastal Impacts: What Water-users Think of Marine Renewables and Why. *Ocean & Coastal Management*, 99, pp. 63–71.

37 Eardley, C. S., and Conway, F. (2011) n. 34.

38 Stokes, C. (2015) *Coastal Impacts in the Lee of a Wave Energy Site: Waves, Beach Morphology and Water-users (Wave Hub, Cornwall, UK)*. Plymouth, UK: University of Plymouth.

39 Ascoop, J., et al. (2012) Stakeholder Input into the Design of the Atlantic Marine Energy Test Site. *4th International Conference on Ocean Energy*. Dublin.

40 Electricity Supply Board International (2011) *Atlantic Marine Energy Test Site: Environmental Impact Statement*. Dublin: ESB International.

41 Electricity Supply Board International (2011) n. 40.

42 Consisting of DONG Energy, RES and B9 Energy Offshore.

43 Yates, K. L., and Schoeman, D. S. (2013).

44 Alexander, K. A., Potts, T., and Wilding, T. A. (2013) n. 15.

45 de Groot, J. (2015) *Attitudes Towards Marine Energy: Understanding the Values in Faculty of Science and Environment*. Plymouth, UK: University of Plymouth.

46 Campbell, M. S., et al. (2014) Mapping Fisheries for Marine Spatial Planning: Gear-specific Vessel Monitoring System (VMS), Marine Conservation and Offshore Renewable Energy. *Marine Policy*, 45, pp. 293–300.

47 Wave Hub (2015) *Offshore Renewable Energy*. Available at: www.wavehub.co.uk/about-us/offshore-renewable-energy

48 Willett, J. (2009) *Why Is Cornwall So Poor? Narrative, Perception and Identity*. Unpublished thesis, University of Exeter.

49 Mills, B., and Cummins, A. (2013) *The Economic Impact of Domestic Surfing on the United Kingdom*. St Agnes, Cornwall: Surfers Against Sewage.

50 SAS (2009) n. 29.

51 Nelsen, C., Cummins, A., and Tagholm, H. (2013) n. 31.
52 DECC (2009) *UK Offshore Energy Strategic Environmental Assessment: Future Leasing for Offshore Windfarms and Licensing for Offshore Oil and Gas and Gas Storage: Environmental Report: Appendix 3h – Other Users and Material Assets (Infrastructure, Other Natural Resources)*. London: Department of Energy and Climate Change.
53 SAS (2009) n. 29.
54 Stokes, C., et al. (2014b) Coastal Impacts of Marine Renewables: Perception of Breaker Characteristics by Beach Water Users. *Journal of Coastal Research*, 70, pp. 389–394.
55 Stokes, C., et al. (2014a) n. 36.
56 SAS (2009) n. 29.
57 Ibid.
58 SAS (2009) n. 29; Millar, D. L., Smith, H. C. M., and Reeve, D. E. (2007) Modelling Analysis of the Sensitivity of Shoreline Change to a Wave Farm. *Ocean Engineering*, 34(5–6), pp. 884–901.
59 Mills, B., and Cummins, A. (2013) n. 49.
60 SAS (2009) n. 29.
61 Stokes, C., et al. (2014b) n. 54.
62 Millar, D. L., Smith, H. C. M., and Reeve, D. E. (2007) n. 58.
63 Ibid.
64 McLachlan, C. (2009) 'You Don't Do a Chemistry Experiment in Your Best China': Symbolic Interpretations of Place and Technology in a Wave Energy Case. *Energy Policy*, 37, pp. 5342–5350.
65 Stokes, C., et al. (2014a) n. 36; Stokes, C., et al. (2014c) Perceptions of the Inshore Wave Resource by Beach Water-Users in the Lee of Wave Hub. *Proceedings of the 2nd International Conference on Environmental Interactions of Marine Renewable Energy Technologies*. Stornoway, Scotland.
66 Stokes, C., et al. (2014a) n. 36.
67 Ibid.
68 Ibid.
69 Stokes, C., et al. (2014b) n. 54.
70 Stokes, C., et al. (2014a) n. 36.
71 Ibid.
72 Ibid; Smith, H. C. M., Pearce, C., and Millar, D. (2012) Further Analysis of Change in Nearshore Wave Climate Due to an Offshore Wave Farm: An Enhanced Case Study for the Wave Hub Site. *Renewable Energy*, 40(1), pp. 51–64.
73 SAS (2009) n. 29.
74 Surfers Against Sewage (2014) *Protect Our Waves – Offshore Renewables*. Available at: www.protectourwaves.org.uk/offshore-renewables.php
75 Surfers Against Sewage (2014) n. 74.
76 Ibid.
77 Dadswell, M. J., and Rulifson, R. A. (1994) Macrotidal Estuaries: A Region of Collision Between Migratory Marine Animals and Tidal Power Development. *Biological Journal of the Linnean Society*, 51(2), pp. 93–113.
78 Tollit, D., et al. (2011) Detection of Marine Mammals and Effects Monitoring at the NSPI (OpenHydro) Turbine Site in the Minas Passage During 2010. *Report to Fundy Ocean Research Centre for Energy*. Available at: https://tethys.pnnl.gov/sites/default/files/publications/Detection_of_Marine_Mammals_at_NSPI.pdf
79 Tollit, D., et al. (2011) n. 78.
80 Fisheries and Oceans Canada (2016) *Review of the Environmental Effects Monitoring Program for the Fundy Tidal Energy Projects*. Canadian Science Advisory Secretariat. Available at: www.fundyforce.ca/wp-content/uploads/2012/05/DFO-Science-Report-2016.pdf

81 Williams, C., and Ward, R. (2016) Bay of Fundy Tidal Power Experiment Approved by Nova Scotia Government. *Canadian Broadcasting Corporation.* Available at: www.cbc.ca/news/canada/nova-scotia/bay-fundy-tidal-project-approved-1.3643256

82 Ward, R. (2016) Bay of Fundy Tidal Project 'Transparency' Questioned. *Canadian Broadcasting Corporation.* Available at: www.cbc.ca/news/canada/nova-scotia/cape-sharp-tidal-power-turbine-bay-of-fundy-fisheries-oceans-1.3662390

83 McMillan, E. (2016) Bay of Fundy Inshore Fishermen Want Court to Overturn Tidal Turbine Approval. *Canadian Broadcasting Corporation.* Available at: www.cbc.ca/news/canada/nova-scotia/bay-of-fundy-inshore-fishermen-wants-judicial-review-1.3697080

84 McMillan, E. (2016) n. 83.

85 Shen, H., et al. (2016) Estimating the Probability of Fish Encountering a Marine Hydrokinetic Device. *Renewable Energy,* 97, pp. 746–756.

86 Shen, H., et al. (2016) n. 86.

87 Conway, F., et al. (2010) n. 10; Devine-Wright, P. (2009) n. 10.

88 de Groot, J. (2015) n. 45.

Chapter 11

Community benefits schemes

Fair shares or token gestures?

Sandy Kerr and Stephanie Weir

1 Introduction

Community benefits in the form of direct payments to local communities are an increasingly common feature of onshore renewable energy. For example, in the UK, there is an expectation of £5,000/MW annual payment from onshore wind projects. These community benefit payments are separate and additional to more general trickle-down economic benefits associated with the development process. In the maritime environment, there is some precedent in case of offshore wind. Benefit payments are now being recommended for other forms of energy development that are perceived to have an impact on local communities (e.g. nuclear power and fracking).

While expectations of community payments are increasing, the language surrounding these transactions is unclear and occasionally contradictory. Arrangements are variously described as 'compensation' or 'goodwill payments'. In the UK, the Government has been cautious of using the word 'compensation', as this could be interpreted as implying a legal right to payments because of some damage or tort. In the normal course of things, private enterprise is not expected to make community payments for undertaking routine, legally permissible economic activity. There is clearly something qualitatively different about renewable energy creating the expectation of additional payments. Communities may have no legal right to demand compensation. However, financial payments *are* being made; and this clearly happens for a reason. Communities may expect payments in response to a perception that private profit is made at the expense of the public good. From the developer's perspective, payments are used to secure some form of benefit or risk reduction. The only alternative explanation is that these are acts of pure altruism, which seems unlikely.

Trickle-down benefits are benefits associated with increased local employment, supply chain expenditure and their respective economic multipliers. These benefit streams are frequently cited by industry and government as part of the "value proposition" used to justify financial support.[1] Many academic and industry studies attempt to value this impact.[2] These benefits are the concern of conventional economics. By contrast, direct community payments, in explicit or tacit recognition of impact on the public good, are not routine market behaviour. Furthermore, these arrangements require the development of new institutional frameworks and legal or quasi-legal processes.

In practice, community payments, as well as the institutions and norms that govern them, vary significantly from situation to situation. The existence of benefit payments and their derivation is not, or at least not exclusively, a matter of economics. This chapter shall explore the issue of community benefits, why they vary from situation to situation, and what the prospects are for community benefit payments in the context of marine renewable energy.

2 Power relations in the energy sector: Gods, Titans and Mortals

Increasingly we are encouraged to think about energy systems in the context of neo-liberal economics and market efficiency. According to this world view, the combined forces of demand and supply and the disembodied invisible hand of the free market allocate resources efficiently. The role of government and its energy policy becomes that of a referee, intervening only to ensure fair play according to the rules of the perfectly competitive market.[3] In theory, individual producers and consumers are powerless, forced to accept market prices determined by the interactions of an infinite number of buyers and sellers.

Jasper[4] uses a radically different metaphor to explain the development of the nuclear power industry. He compares the nuclear sector to the myth of Prometheus, describing a world of Gods (governments), Titans (the power sector) and Mortals (private citizens).[5] Governments are godlike lawmakers with ultimate control over the sector; a small number of titanic organizations (state or private) control the technology and physical resources. Mortals are kept ignorant and largely powerless, buffeted by the ever changing alliances and conflicts of the gods and titans. Jasper suggested that these relations are the key drivers of change rather than disembodied market forces. It is easy to extend this analogy to other sectors. In his classic text, *The Prize*, Daniel Yergin (1990) describes the use of diplomatic, economic and military power to secure access to oil. In the UK, the miners' strike of 1984–1985 is recognised as a pivotal moment in national politics, a power struggle between a right-wing government and Marxist trade unions (Gods and Titans). This conflict paved the way for major change in the energy sector. This included the wholesale privatisation of the energy sector, arguably 30 years of privatisation across the UK public sector, the collapse of the UK coal industry, and the widespread adoption of gas for electricity generation in the 1990s.

Clearly, markets have a significant role to play in the day-to-day operation of the electricity sector. However, energy markets are far from the perfect market ideal of economic textbooks. In the UK, only six companies control approximately 95% of energy supply and 75% of generation.[6] In this context, markets are simply the stages upon which power relations are played out. Other examples abound. The Californian electricity crisis of 2000–2001 saw blackouts for

1.5 million customers. The Texas energy consortium and electricity generator Enron manipulated markets, causing an 800% rise in wholesale electricity prices. This forced the State of California to bail out bankrupt utilities at a cost of US$40–45 billion. Returning to the UK, the recent settlement of a £92.50/MW guaranteed price for electricity from the proposed Hinkley Point nuclear power station was determined through a stand-off between EDF Energy and the UK Government rather than efficient market pricing. These defining moments are contests between Gods and Titans. However, Mortals do occasionally wield power. In 1997, Shell, an oil industry Titan, was forced to abandon its planned deep sea disposal of the Brent Spar oil storage facility because of public protest.[7] Similarly, Germany's rejection of nuclear power and transition to green-friendly distributed generation, the Energiewende, was a response to public unrest after the 1986 Chernobyl disaster.

The issue of community benefits may seem far removed from these events described. However, power relations, as much as markets, determine what happens in the energy sector. This chapter will demonstrate how existing benefit arrangements vary widely from location to location and that it is possible to understand the differences in terms of power relations. This understanding can also help us predict what will happen in emerging scenarios, like the future development of marine energy.

3 Understanding power

Power plays an important role within benefit scenarios, and thus it is important to understand what we mean by 'power' and how it manifests. In a review of the theoretical setting of power, the discourse predominantly rests in the ideas of conflict and domination, informed by the likes of Machiavelli's *The Prince* and Hobbes' *Leviathan*.[8] This concept of 'power as domination' is consistent with the Marxist interpretation, where power is explained as a system of oppression over the powerless and is possessed by the powerful, the repressive hypothesis. In this reading, power is harmful, as are those who wield it. Contrary to this, Foucault approaches power not with the understanding that individuals possess power but rather that power lies within the interactions among individuals.[9] Existing as a relationship that pervades all social organizations, power cannot ever be absolute; instead, it is positioned between actors, usually unevenly. Our interpretation of power can thus move from 'power as domination' to 'power as capacity'.[10] Power in this sense is not necessarily dangerous nor entirely positive; as Galbraith states, "Power can be socially malign; it is also socially essential".[11] Using this interpretation of power, we can understand power in the context of the energy sector as a necessary, sometimes inconsistent and routinely asymmetrical element of the compensation process. This asymmetry, however, does not negate the 'fair price' negotiated between recognisably unequal parties.[12]

This provides us with an understanding of the epistemological foundations of power but not a practical framework of its forms. Galbraith writes on three categories of power, which can work in parallel or separately.[13] These are namely condign, compensatory and conditioned power, operating via punishment, payment or persuasion respectively. Galbraith argues that there has been a noticeable shift from condign to conditioned and compensatory power, as the most common class of power at play in society. Within a renewables context, these kinds of power are expressed by different means; condign power usually manifests by legal means, conditioned power is expressed through social norms, and compensatory power is usually associated with financial instruments.

4 Community benefits and onshore wind

Scotland has a long history of wind energy development and consequently provides a useful timeline of the eventual establishment of near-statutory benefits payments to host communities. The country boasts an impressive record of onshore wind developments, with 5,520 MW of installed capacity and a further 3.9 GW consented for future projects (as of March 2016).[14] These developments require consent by way of planning permission. Establishing the planning process for onshore wind developments

Planning permission is garnered either from the (LA) for developments of less than 50 MW or the Scottish Government for larger developments over 50 MW, which require S36 consent (referencing Section 36 of the Electricity Act 1989). This system of consent permits host communities some say in the outcome of planning queries, particularly in situations where there is strong opposition over the potential negative effects of a development. Benefits payments have consequently evolved as a method of easing public concerns and compensating for a loss of a communal good, although official guidelines maintain that payments continue to be voluntary and not a 'bribe' to aid the planning process.

Despite this assertion of the voluntariness of community benefits, local authorities and the central government have sanctioned non-statutory minimum thresholds for payments from wind developments. Whilst the initial benefits to host communities were usually one-off payments, guidelines now state benefits should be in the form of annual payments, normally calculated in relation to project capacity. Novar Wind Farm, originally developed by National Wind Power Corporation Ltd, was one of the first large-scale wind farms in Scotland, having been built in 1997. Prior to building, the developer also agreed on an annual payment to a community fund of £1,000/MW per year, becoming one of the first wind developments to offer annual payments as a benefit arrangement.[15]

Following this example, the Highland Council suggested an increase in annual payment to £5,000 per installed megawatt, a figure that has been replicated across the industry in both Scotland and England.[16] The current state of community benefits can be found via the public registries: Local Energy Scotland (LES) and English Register of Community Benefits and Engagement (ERCBE). Payments such as these continue to be a form of 'goodwill contribution', yet there is a high level of expectation established in the LA and governmental policies, signifying what could be termed a shadow contract.

5 Offshore wind in the UK

The UK is the world leader in terms of offshore wind deployment, with approximately 5 GW of installed capacity. The largest single development is the 630-MW London Array. The planning processes, which control offshore development, are highly centralised. Local authorities in the UK are consulted about proposed developments; however, the governments in London and Edinburgh are the consenting bodies for English and Scottish waters respectively. In addition to planning permission, developers must secure a seabed lease from the Crown Estate,[17] which manages the seabed around the UK.

While offshore wind technology is to a large extent the same as onshore wind, different legal and institutional regimes have resulted in a very different community benefit arrangements. On land, the industry bodies, local and national governments, and communities have coalesced around an expected payment of £5,000/MW per annum. Offshore, the picture is very different; some developers make no payments at all. The arrangements that do exist are ad hoc and heterogeneous in nature (see Table 11.1).

Benefits schemes associated with offshore wind include donations to wildlife projects, schools, visitor centres, seaside walkways, community centres and lifeboats, as well as student bursary schemes and general 'good causes'. While these payments may be significant, they are both quantitatively and qualitatively different from those related to onshore wind. Despite calls for a standardised regime with the same level of payment as onshore wind,[19] none of these offshore benefit packages are explicitly linked to capacity with annual payments. Furthermore, in most cases, the developer selects the causes and projects that will receive funding. Benefit payments from offshore wind are best understood in the context of corporate social responsibility (CSR), undertaken in order to maintain community goodwill. CSR activity is a form of risk management, which companies undertake to protect their "social license to operate".[20] In the absence of any compulsion, companies have the power to decide how much to invest in CSR, and which

Table 11.1 Community benefit arrangements from UK offshore wind projects

	Annual community fund	One-off community payment	Bursary or education scheme	One-off in-kind payment	No benefits package
Number of projects	7	5	3	8	6

Source: Data correct as of September 2016 (Kerr et al., 2017).[18]

activities to support. Conversely, communities are in a relatively powerless position, able only to respond to ad hoc gifts offered as part of a CRS package. This is substantively different from the situation with onshore wind.

6 The land/sea divide

Steven Jay described the emergence of marine spatial planning (MSP) as something "built at sea", a response to a unique set of circumstances and something very different to terrestrial planning.[21] Unlike land, the sea is a multi-use environment based on the exercise of common rights of access rather than private property rights. This partly explains the emergence of centralised decision making based around rules rather than use zoning. Given that the institutional landscape within which renewable energy operates is significantly different from land,[22] it is perhaps unsurprising that benefits schemes and the motives behind them are equally different. Kerr et al. discuss this divergence further, concluding that it is difficult to imagine an integrated land/sea planning system.[23] Furthermore, centrally controlled MSP and more locally orientated terrestrial planning create the potential for conflict between the two systems.

This issue is particularly significant in the context of marine energy. It is just possible to imagine an energy development that is entirely marine and without terrestrial component: for example, an offshore energy complex (e.g. wind), beyond the exclusive economic zone of any state, producing a transportable fuel (e.g. hydrogen), with limited physical or institutional ties to a specific onshore location. However, the vast majority of proposed marine energy projects will have a hard, physical connection to the land. Offshore wind may be tens of kilometres offshore, with various options for landfall. However, the location of resources and cost drivers mean that most wave and tidal technology deployment is necessarily close to shore. Some technologies are highly transboundary. Aquamarine Power's Oyster wave device pumped high-pressure water ashore with electrical power take-off on land. Such proximity to land greatly reduces the landfall options available to developers. This creates clear potential for tension between terrestrial and marine planning regimes with differing priorities, processes and constituencies. It creates the opportunity for terrestrial and maritime regimes (regulatory and market) to be used as leverage against one another (e.g. securing one permission first) or even one holding the other to ransom.

7 Community ownership

Community ownership of land creates an interesting alternative to the direct payment of community benefits by developers. It also has interesting implications for marine energy developments that straddle the land/sea divide. In Scotland, community ownership of land is actively promoted as a means of redressing historical inequalities in patterns of landownership.[24] Increasingly, community groups are turning to renewable energy as a means of generating income from newly acquired land. This income can come in the form of profits from community-owned projects or rent where communities lease land to developers. In Scotland, public sector support has helped communities develop the capacity to deliver projects. Projects range in scale from micro renewables for direct use to large fully commercial projects.[25] The process of delivering these projects and the income they generate brings additional social benefits, breaking a negative culture of external dependency and empowering local communities.[26]

We have already noted that marine energy projects generally require a landfall. Where offshore developers have a choice of landfalls, they may have the opportunity to negotiate a price with two or more landowners. If, however, there is only one monopoly landowner, then a game of brinksmanship results. If the developer has no alternatives, the landowner can capture a large percentage of the 'economic rent' generated by the project. If the land is in community ownership, the community can find itself in a powerful enough position to maximise financial benefits.

In 2007, the 4-MW-capacity Siadar Wave Power Station energy project, promoted by Voith, received planning consent from the Scottish Government. This project was located off the island of Lewis in the Scottish Western Isles and immediately adjacent to the Galson Estate. The 23,000-hectare Galson Estate was the subject of a community buyout in early 2007. Through a process of negotiation, the Galson Estate's management trust secured a levy of £2/MWh on future electricity exports from the Siadar project.[27] This equates to approximately £6,000/MW of installed capacity per annum, or £24,000 from the 4-MW capacity project at an efficiency rate of 35%.[28] A larger 40-MW project was subsequently planned by Aquamarine Power Ltd. An agreement was negotiated; however, the details remain 'commercial in confidence'. In the end, both the Siadar and Aquamarine projects failed to proceed due to financial and technical difficulties. However, the case illustrates several points: (1) rights in land can be used as leverage against offshore developments; (2) the outcome is radically different from the CSR payments associated with offshore wind projects; and (3) community control over resources significantly shifts the balance of power towards the community.

8 A typology of benefit arrangements

Kerr, Johnson and Weir established a typology that demonstrates the diversity of potential benefits arrangements offered by renewables developments.[29] The typology incorporates a range of important facets used to establish community benefits but concentrates primarily on the power relations that exist between communities and developers (see Table 11.2). Through this understanding of benefits arrangements, Kerr et al. forgo the traditional view of simply balancing costs and benefits, and account for the importance of complex social networks and notions of acceptance.[30]

Table 11.2 A typology of community benefit arrangements

	[1] CSR benefits package	*[2] Benefits schemes*	*[3] Fair shares*
What is the balance of power?	Developer has power to decide	An intermediary acts, with tacit/ assumed support of community	Community can demand payment
What is level of compulsion?	None	No legal obligation *De facto* requirement	High Legal obligations
What is the motivation?	Maintaining a "*social license to operate*"	Removing obstacles to consent	Agreement essential for development to proceed
What is the role of property and property rights?	Irrelevant Separate from any payment decisions	Quasi property rights provide leverage	Critical Property rights are exchanged
What is the nature of the institutional framework?	Social norms	Sectoral norms and stakeholder expectations	Legal
What is the basis of the calculation?	Internal budget allocation	Standardised payments	Directly linked to the profitability of the development

With developer ⟵ *Locus of power* ⟶ With community

[1]

It was conjectured that the type of resulting arrangement is usually depend-ent on where the locus of power lies on this axis. In instances when the power lies closer to the community, the benefits arrangement is more likely to be long term and considerable. The power lies most closely to the community when the communities collectively hold ownership of land. When this occurs, rent can be extracted from developers for the use of the land, in an example of Coasian bargaining.

Moving further along the axis, power will lie closer to the developer, and benefits are arranged via social norms and accepted practice, given that they are not legally required. In these instances, benefit arrangements are aided by the planning permission process via local authorities, which ultimately create the shadow contracts between developers and communities. Arrangements such as these occur in industries like onshore wind, as previously described, where planning consent in the UK creates an exchange of quasi-property rights.

On the end of the scale where the locus of power is situated even further towards the developer, community benefits tend to exist to satisfy corporate social responsibility, and so payments are predominantly at the discretion of the developer. This situation usually occurs when property rights are not featured within the negotiations and no social norms are established. This is the case for the offshore wind industry in the UK, where a quick review found that many off-shore wind projects offered either no benefits at all or one-off, in-kind payments.

9 Thinking forward

This overview of onshore and offshore benefits arrangements has provided an insight into the differences that exist between the two. It is clear that benefit arrangements will vary from situation to situation depending on the balance of power and the prevailing institutional arrangements. So, if governments are minded to encourage community benefits, what can be done to advance and improve the situation that is occurring offshore?

9.1 Emerging systems

New institutional systems are emerging in the energy sector, which seek to pro-vide community benefits based on fairer rates. For example, the Nova Scotia Community Feed-in Tariff (COMFIT) was a Canadian Government–backed programme that rewarded community-financed renewable developments with a steady rate per kilowatt-hour, paid by privately owned Nova Scotia Power. The rates are respective of the type of technology and were established with consider-ation of cost. Consequently, small-scale tidal developments garnered the greatest rate return of 65.2¢/ kW h, whereas owners of terrestrial projects like wind and biomass were paid considerably less. This programme offers a different method for constructing community benefits payments. As opposed to the CSR-satisfying benefits from offshore wind, COMFIT's rates have been established by effective consideration of all the costs involved and an understanding that offshore renew-ables require far greater investment and upkeep.

9.2 Market solutions

In attempting to bridge the inequality between onshore and offshore develop-ments, market solutions could be employed. Utilising the market in environ-mental management has long been a method of reducing negative externalities. Property rights have routinely been imparted onto natural resources, transitioning them from open access as a means of reducing or eliminating externalities and creating markets for exploitation. Additionally, the market is partially used in situations where community ownership of land has allowed Coasian bargaining, due to the well defined property rights allowing for small and private negotiations.

Quasi property rights already exist at sea via the planning process. Developers are given consent to effectively occupy a space that can then no longer be freely utilised by other industries. However, this process does not allow for the same negotiations as with formally defined property rights, as on land. Redistributing property rights at sea away from state control could mean a shift in the perception of the sea being unmarketable or unownable; rather, community ownership of nearshore areas could hypothetically allow more transparent negotiations with the power skewed towards the community. However, similarly, redistribution could also translate to selling areas of the seabed to private corporations, which would have the opposite effect on negotiations.

9.3 Institutional solutions

Institutional solutions, by way of statutory regulations made by authorities, could be key in shifting the accepted norm of benefits payments. There are currently very few institutions that enforce statutory regulations on community benefits from offshore developments, despite the evidence that adjustments in governmental policy can reduce potential opposition (Cowell et al. 2011). In Denmark, regulations ascribe responsibility to developers to offer local communities ownership of at least 20% of both onshore and nearshore wind developments.[31] The policy, part of the Promotion of Renewable Energy Act (2009), has promoted community and cooperative ownership of offshore turbines in a country where there is already a high incidence of community ownership of onshore developments.[32]

Similarly, in Massachusetts (USA), a compulsory programme of community benefit arrangements resulted directly from community petitioning. The Bureau of Ocean Energy Management (BOEM), a federal department of the US Government, introduced the "legally binding contract" for community benefits payments within a leasing notice for their offshore sites.[33]

9.4 Defining 'community'

Another challenge for the future is to understand exactly what 'community' means. Defining the term 'community' has been identified as one of the principal challenges in the establishment of community benefits.[34] Historically, benefits have been arranged for 'the affected community' dependent on geographical location, also termed "communities of locality".[35] However, spatial boundaries on localities are usually vague and complex. Additionally, communities are not homogeneous groupings with shared opinions. Another definition is that of 'communities of interest', groups of people who share a common outlook on various social facets of life. Even through this understanding, forming the limits of a community of interest is difficult.

Establishing the extent of a community of either locality or interest is significantly more complicated when the development is placed offshore, where there is "increased spatial separation" between the local community and the infrastructure.[36] When visual impact is considered, for example, it is possible that more than just the closest community will be negatively affected by an offshore wind development. Despite this, communities are routinely defined by "spatial determinants" for both onshore and offshore developments.[37] Often, community projects were begun prior to any formal definition of community, which only serves to exacerbate the problem of benefits distribution.[38]

10 Conclusion

Community benefits schemes from renewables developments are in no way uniform. Literature has repeatedly shown that these arrangements can take many forms. The typology provided by Kerr et al. explains that this multiplicity of arrangements is a response to the specific institutional circumstances and power relations at work within community benefits negotiations.[39] The typology provides a spectrum of arrangements dependent on these institutional factors. Arrangements can range from voluntary corporate social responsibility, as evidenced in the case of offshore wind, to fair shares programmes, where communities wield heightened power through land ownership.

Should policymakers wish to encourage benefits payments, policy needs to be introduced that adequately responds to the highly specific nature of each situation. Moreover, statutory approaches for securing payments may be required where communities have limited leverage. They have been proven to be effective in countries like Denmark, prompting developers to be increasingly compelled to provide a variety of benefits arrangements because of legal obligation.

For offshore developments, evidence suggests that benefits arrangements are best characterised as CSR activity. These are not standardised or negotiated and generally take the form of gifts to good causes. The typology has shown this is due to a highly centralised planning process, a lack of an exchange of property rights, and the community having very little power or influence. If communities were given the opportunity to control their adjacent land or even sea space, different outcomes may result. This observation is of particular resonance in the UK, where plans are currently in place to devolve powers of administration over the territorial seabed from the Crown Estate to the Scottish Government. Further devolution of rights to a local level could allow communities to extract rent from developments within their local sea space. There are already precedents for such arrangements. Shetland, for example, saw a transfer of marine rights prior to major oil and gas developments in the region.[40]

However, the mechanisms for the transfer of rights in resources are still unclear, particularly where the state appears to want to maintain control. This is evidenced in the ongoing tension between localism and centralisation that is occurring within the current policy landscape. In the UK over the past 25 years, there has been a process of localism in land planning, with the devolution of powers away from the centre.[41] Localism manifests itself in other forms, including land reform and the reassertion of traditional rights.[42] Whilst the Scottish Government continues to strongly advocate for empowering communities, it maintains a highly centralised process for marine planning, which does not allow for much community input. More generally, the need for national policies and strategies regarding climate change and energy infrastructure can result in national and local priorities at odds with each other.

Additionally, a number of unanswered questions within the community benefits discourse still need to be addressed before policies can be optimised. Particularly challenging is the question of defining 'community'. As shown, oftentimes establishing boundaries for a community to receive benefits payments is challenging, even more so when the development is placed at sea. Policymakers must work to forge a more encompassing understanding of 'community' before benefits arrangements can work most efficiently.

Notes

1 Gardiner Pinfold (2015) *Value Proposition for Tidal Energy Development in Nova Scotia, Atlantic Canada and Canada: Offshore Energy Research Association.* Halifax, Nova Scotia: OERA. Available at: www.oera.ca/
2 Gilmartin, M., and Allan, G. (2015) Regional Employment Impacts of Marine Energy in the Scottish Economy: A General Equilibrium Approach. *Regional Studies*, 49(2), pp. 337–355.
3 (1) No one player can influence prices in the market; (2) homogeneous product; (3) perfect information about products and prices; (4) free entry and exit from market.
4 Jasper, J. (1992) Gods, Titans and Mortals: Patterns of State Involvement in Nuclear Development. *Energy Policy*, 20(7), pp. 653–659.
5 Ibid.
6 (1) British Gas, (2) EDF Energy, (3) npower, (4) E.ON UK, (5) Scottish Power, and (6) SSE.
7 Side, J. (1997) The Future of North Sea Oil Industry Abandonment in the Light of the Brent Spar Decision. *Marine Policy*, 21(1), pp. 42–52.
8 Karlberg, M. (2005) The Power of Discourse and the Discourse of Power: Pursuing Peace Through Discourse Intervention. *International Journal of Peace Studies*, 10(1), pp. 1–23.
9 Foucault, M. (1996) *Live Interviews, 1961–1984.* Hochroth, L. and Johnston, J. (trans.). New York: Semiotext; Mills, S. (2003) *Michel Foucault.* London: Routledge.
10 Karlberg, M. (2005) n. 8.
11 Galbraith, J. K. (1983) *The Anatomy of Power.* London: Corgi, p. 13.
12 Whyte, J. (2013) Is Revolution Desirable? Michel Foucault on Revolution, Neoliberalism and Rights. In: Golder, B. (ed.) *Re-reading Foucault: On Law Power and Rights.* London: Routledge, pp. 207–229.
13 Galbraith, J. K. (1983) n. 11.
14 Scottish Renewables (2016) *Renewables in Numbers.* Available at: www.scottish renewables.com

15 Centre for Sustainable Energy (CSE) (2009) *Delivering Community Benefits from Wind Energy Development: A Toolkit – A Report for the Renewables Advisory Board*. Bristol, UK: CSE.

16 Highland Council (2003) *Community Benefit from Renewable Energy Projects*. Report by the Chief Executive. Inverness, Scotland: Highland Council.

17 The responsibilities of the Crown Estate in Scotland are currently (2017) being devolved to the Scottish Government.

18 Kerr, S., Johnson, K. and Weir, S. (2017) Understanding Community Benefits Payments from Renewable Energy Developments. *Energy Policy*, 105, pp. 202–211, n. 18.

19 Highland Council (2013) *Guidance on the Application of the Highland Council Community Benefit Policy for Communities and Developers*. Inverness, Scotland: Highland Council.

20 Hall, N., et al. (2015) Social Licence to Operate: Understanding How a Concept Has Been Translated into Practice in Energy Industries. *Journal of Cleaner Production*, 86(1), pp. 301–310.

21 Jay, S. (2010) Built at Sea. *Town Planning Review*, 81(2), pp. 173–191.

22 Todd, P. (2012) Marine Renewable Energy and Public Rights. *Marine Policy*, 36, pp. 667–672.

23 Kerr, S., Johnson, K., and Side, J. (2014) Planning at the Edge: Integrating Across the Land Sea Divide. *Marine Policy*, 47, pp. 118–125.

24 LRRG (2014) *The Land of Scotland and the Common Good*. Land Reform Review Group. Available at: www.scotland.gov.uk/Resource/0045/00451597.pdf

25 As of March 2017, the largest community project in the UK is a 9-MW wind project managed by Point and Sandwich Development Trust in the Western Isles. Available at: www.pointandsandwick.co.uk/

26 Rennie, F., and Billing, S. L. (2015) Changing Community Perceptions of Sustainable Rural Development in Scotland. *Journal of Rural Community Development*, 10(2), pp. 35–46.

27 Kerr, S., Johnson, K., and Weir, S. (2017) Understanding Community Benefits Payments from Renewable Energy Developments. *Energy Policy*, 105, pp. 202–211.

28 This assumes a device efficiency of 35%.

29 Kerr, S., Johnson, K., and Weir, S. (2017) n. 26.

30 Ibid.

31 Anker, H. T., and Jørgensen, M. L. (2015) *Mapping of the Legal Framework for Siting of Wind Turbines- Denmark*. Frederiksberg: Department of Food and Resource Economics, University of Copenhagen (IFRO Report No. 239).

32 Mendonça, M., Lacey, S., and Hvelplund, F. (2009) Stability, Participation and Transparency in Renewable Energy Policy: Lessons from Denmark and the United States. *Policy and Society*, 27(4), pp. 379–398.

33 BOEM (2014) *Massachusetts Proposed Sales Notice*. Available at: www.boem.gov/ BOEM-MA-Auction-Seminar-PSN-Overview-Presentation/

34 Munday, M., Bristow, G., and Cowell, R. (2011) Wind Farms in Rural Areas: How Far Do Community Benefits from Wind Farms Represent a Local Economic Development Opportunity? *Journal of Rural Studies*, 27(1), pp. 1–12.

35 Department of Trade and Industry (DTI) (2005) *Community Benefits from Wind Power. A Study of UK Practice & Comparison with Leading European Countries – Report to the Renewables Advisory Board and the DTI, 05/1322*. London: DTI.

36 Haggett, C. (2008) Planning, Politics and Public Perception of Offshore Wind Farms. *Journal of Environmental Policy & Planning*, 10(3), pp. 289–306.

37 Rudolph, D., Haggett, C., and Aitken, M. (2014) Community Benefits from Offshore Renewables: Good Practice Review. *ClimateXchange* 9. Available at: www. climatexchange.org.uk/files/7314/2226/8751/Full_Report_-_Community_Benefits_ from_Offshore_Renewables_-_Good_Practice_Review.pdf

38 Walker, G., and Devine-Wright, P. (2008) Community Renewable Energy: What Should It Mean? *Energy Policy*, 36(2), pp. 497–500.
39 Kerr, S., Johnson, K., and Weir, S. (2017) n. 26.
40 Johnson, K. (2013) Shetland: A Model for the Future. *The Shetland Times*, 5 April.
41 Holman, N., and Rydin, Y. (2012) What Can Social Capital Tell Us About Planning Under Localism? *Local Government Studies*, 39(1), pp. 71–88.
42 Kerr, S., et al. (2015) Rights and Ownership in Sea Country: Implications of Marine Renewable Energy for Indigenous and Local Communities. *Marine Policy*, 52, pp. 108–115.

Chapter 12

Consultation in ocean energy development

John Colton, Flaxen Conway, Bouke Wiersma,
Jordan Carlson and Patrick Devine-Wright

1 Introduction

Consultation is increasingly viewed as a critical factor in renewable energy development, including ocean energy (OE) development. Failure to consult the community and other stakeholders and failure to employ the appropriate methods for consultation will, in many instances, impact the successful development and implementation of the OE project. This chapter provides a concise summary of key issues related to the principles and practices of consultation, drawing on renewable energy scholarly literature and specifically OE literature. Key concepts including 'social license' and 'social acceptance' are explored in more depth, as well as the role of consultation in mitigating conflict related to other uses of the marine environment.

A series of three case studies drawn from European and North American contexts illustrate the complexity of consultation and its role as a critical tool and factor in supporting responsible OE development. The role of place meanings and using visual consultation methods at an early stage of project development is explored in Guernsey, a British Crown dependency and one of the Channel Islands, where offshore wind, tidal and wave energy are being considered. In Oregon (USA), consultation has enabled the integration of a grid-connected wave energy testing facility that also supports research. The final case study from Nova Scotia, Canada, emphasizes the role of consultation in the development of strategic planning documents and policies that provide guidance to tidal energy development in the Bay of Fundy. The chapter concludes by summarizing key lessons and themes drawn from the literature and case studies for consideration for future practice and research.

2 What do we mean by 'community consultation'?

'Community consultation' is a vague concept. In its simplest form, community consultation can be considered any form of communication between the proponent of a given development project or policy and the communities that they will affect.[1] Though often conceptualized as a one-way process where developers consult communities on a proposed project, Walker et al. describe consultation as a two-way process that is shaped by both 'sides' and the expectations they may hold of each other.[2] For instance, it has been argued that developers' expectations of a hostile public may lead them to shy away from consultation and carry out only minimal consultation at the latest stage possible.[3]

Moreover, the question of what constitutes a 'community' and/or a 'public' is important yet complex to answer given that the terms are often used interchangeably. Although the term 'publics' is often used to reflect the plurality and diversity of what should not be seen as a homogeneous 'general public' – as is illustrated by the diverse array of stakeholders consulted in Oregon and Nova Scotia, in this chapter the term 'public' is used to refer to a general, non-local population. This contrasts with the term 'community', which is understood here to refer to those individuals and groups directly affected by a localized energy project. The question of whom to consult when practicing consultation is therefore an important one to consider for policymakers, developers and academics alike.

3 Why do we consult the public?

As the general public has become more aware of the environmental and socio-economic consequences of development choices, there has been a general trend of greater scrutiny of proposed projects, including renewable energy installations.[4] This greater scrutiny has led to calls for more public involvement and authority over decision-making processes.[5] In much of the world, the response of governments and private interests has been the introduction of formal, legally founded public consultation processes.[6] Motivations for consultation can vary widely yet several themes emerge with deeper analysis. Table 12.1 presents a summary of many common motivations for consultation including a brief description of each theme.

An analysis of motivations behind consultation does not fully answer the question of why public participation is pursued.[7] Here, Glucker et al.'s analysis of the purposeful rationales of consultation is useful.[8] Consultative rationales can, broadly, be broken into three groups: the instrumental, the substantive, and the normative. Using an instrumental rationale, consultative processes are viewed as a means of providing legitimacy to any project decisions that are made, as well as serving to resolve conflicts between parties.[9] According to substantive rationales, the public should be involved in planning projects in order to make the project outcomes better, through providing local knowledge, experience and feedback.[10] Finally, normative rationales assume that consultation should be pursued on the basis of strengthening democratic capacity in the citizenry.[11] That is, normative rationales see consultation as a means of developing engaged citizens, encouraging social learning and allowing greater influence for local people in decisions that affect them.

Obviously, these rationales suggest a wide divergence in beliefs about the purpose of consultation. This wide variance can be explained by the fact that no two participants in consultation exercises are going to share identical

Table 12.1 Motivations for consulting the public

Theme	Description
Acceptance	Primary goal of consultation: achieve local acceptance. Can and will vary at different levels; e.g. society may believe in renewables, but the market may not support them. Broad sociopolitical support for alternative energy may coexist with local resistance to a given project.
Early contact	Consultation should begin as early as possible in the planning stages. Fosters trust and communication. Allows local concerns and views to be taken into account early on. Allows for protocols to be established regarding what and when information should be shared.
Trust	Absolutely necessary to successful project implementation. Naturally builds very slowly. Can be lost very quickly.
Environmental, noise and visual impacts	Impacts on audiovisual land- or seascapes or environment can cause opposition. Renewables can be seen to either 'fit' in the landscape or not. 'Fit' influences perception as benefit or detriment.
Community benefits or (co-) ownership	Direct benefits a key factor in achieving acceptance of a project. Local (co-)ownership, payments or lowered energy prices as possible methods to create these benefits. Many projects are still developing novel models to attempt benefit sharing.
Equitable and just decision making	Major concerns exist over the fairness of decision making regarding renewables siting. Exacerbated when developers are not 'local'. Participatory processes a method of achieving equity.
Information sharing	Accessible and easily understood information vital.
Endogenous development	Reliance on local resources, workers and people. Incorporation of local beliefs and priorities in planning.
Unknowns	Consequence of novelty of projects. Limits promises of developers and expectations of communities.

rationales for being involved.[12] For a developer, pursuing public participation can be seen as many things – a legal obligation, a burden, a way to maintain public favour, a way to avoid backlash, or a genuine way to attempt to improve their project planning and development approaches.[13] A government's rationales for requiring participation can either be rooted in attempts to create an engaged, democratically empowered public or in the desire to avoid public outcry – and thus get re-elected.[14] The public's rationales for choosing to be involved can vary even more greatly; while some individuals may be interested in how they may profit from a project, others will be concerned with how it will disrupt their way of life or disturb the landscape and aesthetics of places they are attached to.[15]

4 Consultation and renewable energy

Although there remains a great deal of debate over what consultation is and how it ought to be carried out, common themes exist with respect to consultations for renewable energy projects. The central element needed for successful consultation – agreement by nearly all authors – is trust among all parties involved. To generate this trust, it is considered vital that renewable energy proponents approach communities and inform them of the project and their plans at the earliest opportunity in the development process.[16] That said, even when a community trusts the intentions of a project developer, they may not entirely 'approve' of the proposed project. However, a lack of approval does not necessarily constitute a rejection of a project; a community's constituents may accept the presence or necessity of the project, without outright endorsing it or considering it to be positive.[17] In other words, getting a community to accept a project can sometimes mean pursuit of a bare minimum of community tolerance rather than endorsement of the installation.[18] This distinction between local acceptance/tolerance and active local support/local approval of projects must thus be borne in mind when discussing consultation and its goals. Table 12.2 explores themes in consultation and renewable energy projects.

Table 12.2 Key themes in consultation for renewable energy projects

Motivation	Reasoning	Interested parties
Trust building	Establishing a relationship with communities encourages dialogue and understanding.	Developers Regulators Planning authorities Communities
Lessening resistance	Providing information and getting feedback can generate or increase support.	Developers Regulators Planning authorities
Meeting regulatory requirements	In most jurisdictions, consultation is legally mandatory; compliance is vital to positive perception.	Developers Governments Regulators
Avoiding backlash	Perceived failures to 'adequately' consult people can engender resistance to projects and government.	Developers Governments Regulators

Motivation	Reasoning	Interested parties
Involving local industry	Involves creating local jobs and avoiding disruption to local industries.	Developers Local industries Planning authorities
Determining local interests	Allows synergistic approaches to developments, accounting for local desires and priorities.	Developers Planning authorities Communities Local industries
Mitigating risk and environmental impacts	Identify local environmental concerns and issues; inform developers and planners of risks of the project.	Developers Planning authorities Environmental groups Regulators Communities Local government
Public opinion ground-truthing	Local views will vary with location; ascertaining local context is a necessary first step.	Developers Regulators Planning authorities Local government

What is apparent from these themes is that, on the whole, consultation is considered a necessary part of the planning process for any renewable energy project. That said, there remain many disagreements on how to carry out such consultations, who should be involved, and to what extent project proponents should delegate decision-making powers to the public. It is often claimed that consultation can lead to communities seeing greater benefits from projects; however, the reality of such benefits depends heavily upon the means and extent of public participation.[19]

5 Social license and social acceptance

Important concepts emerging in light of discussions centred on public consultation are *social license* and *social acceptance*. The term 'social license' originated two decades ago in the mining industry context where industry consultants emphasized the growing importance of working to meet and address local community needs beyond economic impacts.[20] Underlying the concept of social license are the issues of credibility, legitimacy and trust – all factors deemed largely missing in many development contexts but critical in advancing a social contract between a community and industry and/or government.[21] Social license is intangible and "associated with acceptance, approval, consent, demands, expectations, and reputation".[22] In some cases, it implies working towards greater environmental and social performance in order to meet stakeholder expectations but in doing so may very well exceed regulatory policies associated with a development project. While the concept of social license is vague and not part of any statutory and permitting framework, the issue of addressing broader social, environmental and economic issues with respect to development projects has become a key determinant in a project's success.

'Social acceptance' is the term more normally associated with renewable energy development, especially in the European context where you seldom encounter the term 'social license'. Renewable energy development, its associated polices and technologies have been shaped by the growing importance of the issue of social acceptance.[23] While the concept of social license has been challenging to measure, researchers have explored a number of variables that provide insight into social acceptance.[24]

Some of these variables include attitudes and behaviour, age, socio-economic background, and political views. Place attachment can also be a determining factor in the level of social acceptance for a renewable energy project and has been found, in some instances, to be more influential variable than socio-economic variables[25] as suggested in the Guernsey case study.

A fundamental issue of both social license and social acceptance is trust and that renewable energy projects informed by processes supporting the development of trust are more likely to be accepted by the local community and larger public. The motivations for consulting the public and the themes highlighted in renewable energy development and consultation suggest that, increasingly, renewable energy development projects are being supported by consultation processes that *may* support the development of trust. While a wide range of consultation and engagement tools, processes and strategies exist, what are more important are the principles and values that support the consultation process. These might include, for example, transparency, early and ongoing engagement, and sharing of research data and other important information about the project. Failure to address these important principles and/or to develop meaningful consultation strategies can result in conflict.

6 Conflict and consultation

One of the simplest and eloquent definitions of conflict is the natural tension that arises from differences. Far from being unplumbed or untapped, marine and coastal waters provide a wealth of benefit to humans and the planet. The ocean is a busy place with multiple users that have overlapping objectives that may not be compatible and that, in fact, are competitive.[26] OE is another resource and OE-related testing, research and industrial sites would be an "additional user" that could set the stage for further conflict.

The common reaction or approach to potential conflict is to seek 'mitigation', and, in the marine context, this has translated to offering financial remunerations to traditional users to give up their claims to the contested space. Such pay-offs, however, tend to benefit only the original recipients. Efforts to create longer-term value for the existing users are possible but can succeed only through a collaborative solution-seeking approach. Thus consultation can be viewed as a conflict avoidance strategy.

To avoid conflict and the loss of the social, cultural and economic value associated with existing uses of the sea is optimal. For example, consultation with existing users of ocean space could yield understanding of the tensions that typically arise from existing use overlap. More importantly, however, consultation illuminates the strategies used to lessen or avoid the tensions that arise from that overlap.[27] Further, consultation might benefit from focusing less on the question of where the use or overlap occurs to focusing more on who uses the space or resources and why. It is through these conversations that mutual interests and potential strategies to support them can be identified.

Consultation in OE development requires taking the time to understand users and the reasons for their use. Consultation illuminates how proposed actions would impact current users (and vice versa). Improved understanding and awareness of whether the proposed use would be compatible or incompatible with extensive, and long-standing uses is the first step towards finding ways to lessen the tension that arises from differences.

7 Case studies in ocean energy development

The case studies highlight diverse contexts, OE technologies and consultation processes associated with OE development. The purpose in sharing these cases is to broaden our collective understanding of the contexts in which OE development occurs and the consultation and engagement processes that support this development.

7.1 Public acceptability of offshore renewable energy in Guernsey: using visual methods to investigate local energy deliberations

One case study that offers an example of consulting local communities at an early stage of project development is offered by recent interdisciplinary PhD research by Wiersma.[28] This research explored how the potential future development of ocean energy is made sense of and evaluated by residents of Guernsey, a largely autonomous British island in the English Channel. Guernsey has a significant offshore wind, tidal stream and wave energy resource but no concrete proposals for developing these. As such, it offers a useful case study context to further examine how local communities make sense of and evaluate novel ocean energy technologies. This is important given that the public acceptability of ocean energy is a relatively underdeveloped area of the public acceptability literature.[29]

This research aimed to address two key shortcomings in the existing literature on the public acceptability of local energy projects.[30] The first of these is that this topic is commonly studied by examining public responses to single developments that have recently been proposed by developers.[31] Although this approach has yielded many valuable insights,[32] it overlooks 'upstream' factors that play out at earlier stages in the development process, such as the choice of technology and location for a given energy project. Also, through a subsequent focus on oppositional public responses, the willingness and ability of local communities to contribute constructively to the design of locally supported energy developments has also frequently been overlooked. The existing literature therefore has relatively little to say about how the acceptability of specific energy projects is influenced by choices of technology and site, about which local alternatives may have been more acceptable in a given context, and about whether asking local residents about multiple potential future local energy projects at an early stage can produce more acceptable energy projects.

These questions have two implications that informed this Guernsey case study. First of all, it suggests that it is important to understand how a place is viewed by its residents, as such place meanings may influence how subsequent place change is interpreted locally: *a place-based approach.*[33] Studies adopting this approach have previously found that it is important for technologies to be seen to 'fit' a given place: for instance, technologies seen as 'industrial' are argued not to fit places seen as 'natural',[34] while if a technology is seen as an economic opportunity, it is seen to 'fit' places considered struggling economically.[35] The second implication is that this Guernsey case study is not primarily interested in whether a project that has already been designed and proposed is 'acceptable', but instead explores whether acceptable kinds of renewable energy development can be identified by engaging local communities at an early stage and more broadly with their local energy future. The potential benefits of early public engagement have been described by many scholars; nonetheless, few energy acceptability studies have explored acceptability questions at an early stage.

The research described here consists of three mixed-method studies. Working collaboratively with the Guernsey Government's Renewable Energy Team, it investigated how potential future offshore wind, tidal and wave energy projects were represented by Guernsey residents to threaten, enhance or fit place-related values and meanings associated with Guernsey and its coast and sea. Also, as previous acceptability studies have often been verbal or text based, this research aimed to explore the usefulness of visual research methods in order to further the methodological diversity of this field. In this chapter, only a brief overview of its methods and key findings is given, as a more in-depth account is beyond the scope of this chapter.[36]

7.1.1 Study 1

The first Guernsey study used auto-photography,[37] by asking 28 participants to take up to ten photographs of 'what they value about Guernsey's coast and sea' and subsequently interviewing them about these photographs. The method was chosen as it was expected to enable an in-depth exploration of specifically local ideas about the Guernsey coast, and their impact on how potential changes to this place are made sense of. One common theme throughout the photographs and accompanying stories (which can be found in Wiersma, 2016) was Guernsey's large (up to 10m) tidal range as something that is a very important and valued part of life on the island, making Guernsey a *unique* place. This notion of Guernsey being a highly unique and distinct place was widely returned to in subsequent discussions of the acceptability of different offshore energy sources (offshore wind, wave and tidal energy) by representing tidal energy as a highly acceptable future energy source because it fits with and enhances this distinctiveness. By contrast, offshore wind energy was talked about as "making Guernsey more like everywhere else". Other key themes represented by the highly diverse participant photographs portrayed Guernsey as an island in need of more independence, with a coast valued for its quietness, wildlife, leisure opportunities, tides, natural beauty, and as a space for exploration and utilisation – all of which had implications for local energy acceptability.[38] The use of auto-photography in this study was thus able to open up important place meanings and to use these to understand which offshore energy technologies are considered to fit these meanings.

7.1.2 Study 2

The second Guernsey study further explored the acceptability of future tidal and offshore wind energy projects in Guernsey, with an emphasis on findings suitable locations for such projects. Four deliberative focus groups were held with five or six participants each. They were asked to place one green and one red sticky note on a large A1 map of Guernsey to signify which locations would be most acceptable (green) and least acceptable (red) to them as sites for offshore wind/ tidal energy development. Figure 12.1 shows a map made by one of the groups for offshore wind locations, which clearly shows locations to the north of the island being marked as more acceptable than locations to the west.

Discussions accompanying this task revealed that the chosen location for offshore energy projects is a very important *condition* for support.[39] Even those expressing very supportive views in general talked about certain places where they would be opposed to such projects:

I: In what way would [offshore wind development] be unacceptable to you?
FEMALE PARTICIPANT: If they put it on the Humps [an area known for its wildlife].

This map-based task thus opened up the question of what would be 'the right place' for either offshore wind or tidal energy projects. It revealed that the precise location chosen for such energy developments within a larger locale (Guernsey as a whole, in this case) is very important in developing an acceptable local energy project. This was the case for both offshore wind and tidal energy (although due to the uncertainty surrounding the physical impact of tidal stream turbines, participants found it more difficult to complete this task in relation to tidal energy). The discussions also revealed the multitude of ways of thinking that were used to argue for or against places, suggesting that local residents care about more than

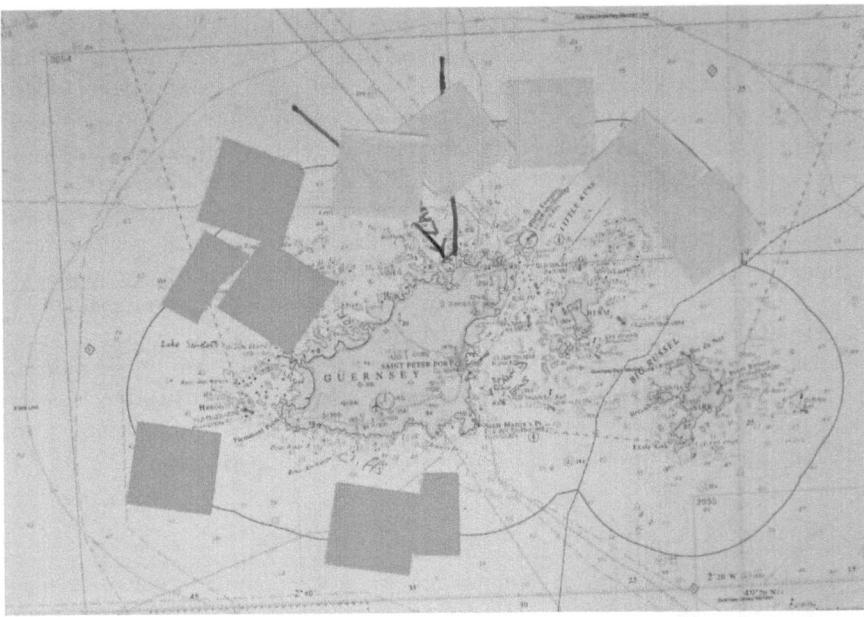

Figure 12.1 Participants' sticky notes representing (un)acceptable sites for off-shore wind development

just the visual impact (a concern frequently foregrounded in the literature). For instance, places seen as already built-up and 'industrial', places where 'others' lived (neighbouring Jersey, for instance), as well as places thought to be financially most viable or lacking in wildlife were represented as acceptable areas for offshore energy development. On the other hand, places with which people have a strong personal bond, places seen as widely popular or places known for their views or sunrises and sunsets were represented as the least acceptable locations for offshore energy projects.

7.1.3 Study 3

The third Guernsey study aimed to further build on these qualitative findings using an island-wide representative sample questionnaire survey (*N* = 468). This study confirmed quantitatively the main conclusions from the first two studies: for instance, that tidal energy is considered much more acceptable than offshore wind energy (and, unlike offshore wind, is supported equally across gender, age, income and education groups).

Importantly, study 3 also asked about the acceptability of three broad zones that may be technically feasible locations for a 30-MW offshore wind project (A, B, C) (Figure 12.2a) or a 30-MW tidal energy project (X, Y, Z) (Figure 12.2b).

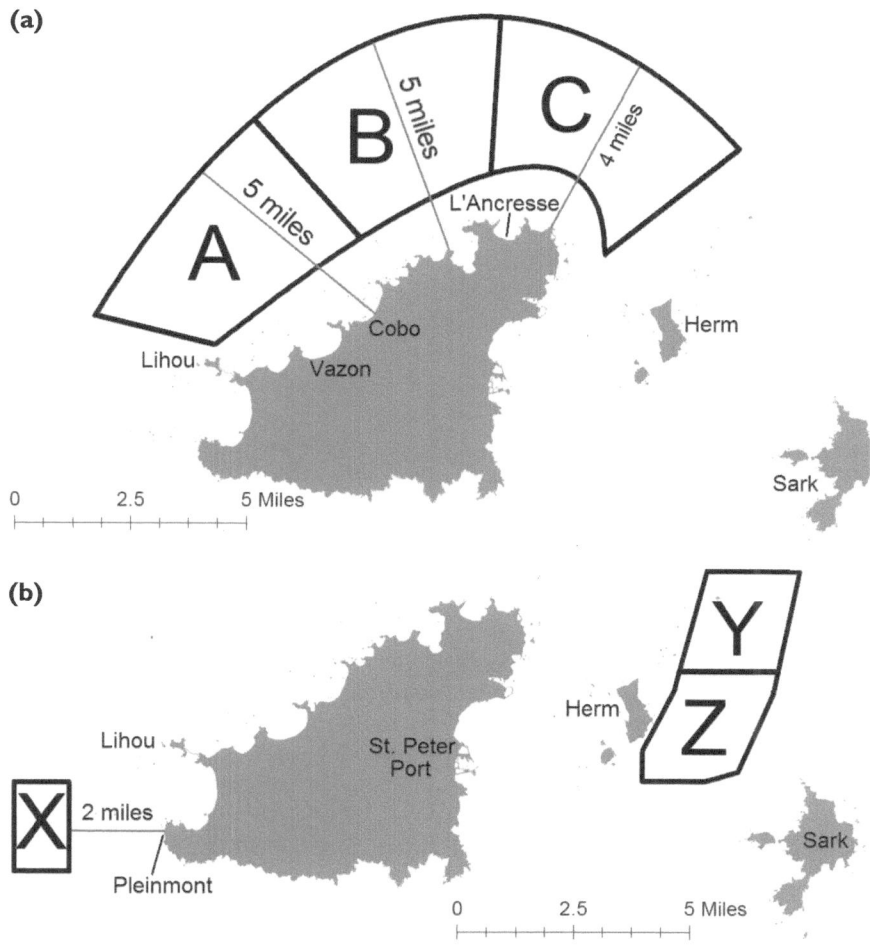

Figure 12.2 Zones technically feasible for (a) offshore wind (A, B, C) and (b) tidal
      energy (X, Y, Z)

This revealed that all of these zones were evaluated significantly differently in
terms of their acceptability as sites for offshore energy development, despite being
sited at a similar distance from the shore. For instance, zone A was supported as
a site for an offshore wind project by only 30%, while zone C was supported by
a small majority (52%). Similarly, zone Z was the least popular zone for a tidal
project (supported by 49%), while 74% supported the same project in zone X.
The importance of finding the right location is illustrated here by the fact that

a well supported technology (tidal) in the 'wrong' place is less widely supported than an unpopular technology (wind) in the 'right' location. Also, these results are particularly significant for tidal energy because the hypothetical tidal energy project was described as being entirely submerged: clearly this does not mean that location is irrelevant and the technology is 'out of sight, out of mind'. Instead, the many different arguments used during the focus groups seem to apply to offshore wind projects as much as to tidal energy projects.

7.1.4 Conclusions

This Guernsey case study suggests that it can be very worthwhile to explore the 'social' suitability of different local energy options at an early stage. If only technical or financial considerations are taken into account, a technology and a site may be selected that from a social point of view simply do not fit (e.g. zone C for offshore wind). Importantly, this was found to be true for both offshore wind energy as well as tidal energy technology – despite obvious differences in aspects like visual impact. This suggests that it is important for future consultations to take public concerns about a technology and the places where it may be sited seriously, as the fit between place and technology representations can be a key influence on the acceptability of ocean energy projects.[40]

This case study suggests that focusing on place meanings can be a helpful way of analysing this fit between representations of both place(s) and technologies. Finally, the visual methods adopted (especially auto-photography) were found to be very valuable tools in going beyond a discussion purely about energy but instead reframing the conversations as being about the future of a place and starting a constructive dialogue about the different potential types of place change that may or may not be acceptable. Also, it was found that using an auto-photography methodology helps in recruiting those with less strong views on renewable energy and is enjoyable for both researcher and participants, contributing to a successful consultation process.

In short, this case study suggests that early-stage consultation using a visual, place-based approach could be one way to both achieve a better academic understanding of the acceptability of local energy projects and to contribute to the development of more acceptable energy development practices in the future.

7.2 Ocean energy development in Oregon

This case study explores the opportunities and some of the barriers in developing wave energy in the Pacifica Northwest of the United States. Oregon has promoted the development of OE off its coast since 2006. This is no surprise as the northeast boundary of the Pacific Ocean has some of the world's best wave energy resources.[41] Oregon has several ports and harbours to fabricate and maintain wave energy installations, an established coastal electrical grid with excess capacity and a nearby population centre with electricity demand.[42] Based on these characteristics, seven areas off the coast of Oregon that could accommodate 100-MW commercial scale sites were identified.[43]

About 42% of Oregon's electricity comes from hydropower, followed by coal (41%), natural gas (10%), nuclear (3%), biomass (3%) and wind and geothermal (1%).[44] In February 2006, Oregon's governor proposed a Renewable Portfolio Standard (RPS) for Oregon that called for 25% of the state's electricity to be generated from renewable sources by 2025. In 2007, the state legislature passed Senate Bill 838 codifying the governor's proposal, and additional bills provided for a variety of renewable incentives and efficiency improvements, including funding for research and development, tax credits for renewable firms and installation of energy-efficient appliances for homeowners. Eligible sources of electricity include those generated from wind, solar, wave and tidal, geothermal, biomass and others accepted under Oregon's rule-making process. Engineers at Oregon State University (OSU) estimated that 100-MW developments of wave energy generators could provide approximately one-half of the renewable energy required under the 25% by 2025 RPS requirement.[45]

Despite the cautious optimism and genuine interest of key players (including local leaders, state and federal agencies, academics and industry leaders), the future of wave energy off Oregon's coast remains uncertain. In many ways, wave and floating wind technology is still in its infancy, much like land wind technology 25 years ago. Several wave energy conversion devices have been developed, but no single technology has been proven superior. In addition, the marine resources managed off Oregon's coast are vast; the near-shore environment is characterised by a variety of habitat types, including rocky intertidal zones, soft sandy bottoms, kelp beds and rocky reef habitats, each habitat supporting many species of invertebrates, fish and marine mammals (some of which are unique to particular habitat types), which underpin commercial and recreational industries. Thus, while the state may be well suited due to an abundant wave resource and supporting coastal infrastructure, there remain potential barriers to implementation of the RPS and the OE industry due to uncertainties about the technology and concerns for impacts on the local environment. While it is essential to pay attention to the ecological impacts of this emerging industry, permitting processes rarely fail on technical or natural science grounds but rather because of a failure to pay attention to human dimension concerns.[46] These uncertainties and concerns ultimately relate to consultation (called public and/or stakeholder engagement in the United States), the desired outcomes of which should be increasing understanding, lessening resistance, involving local industries and considering their interests, avoiding environmental and social impacts, and building trust and fostering public acceptance.

Some initial investments in social science research can inform the design and implementation of consultation. In Oregon, this took the form of the OSU Human Dimensions of Wave Energy research programme in the late 2000s[47] and other research efforts over the next decade in the areas of OE and marine spatial planning.[48] These efforts focused primarily on the human dimension and took a mixed-methods approach (qualitative and quantitative) to studying, analysing and reporting research findings. Most recently, research focused on the importance of consultation with regard to the siting of the Northwest National Marine Renewable Energy Center's grid-connected, open-ocean test facility for full-scale wave energy devices. Specifically, this research investigated whether participants in the siting process understood the process, felt as though they were heard and felt they had an influence on the outcome of the process.[49] While this process was focused on siting a 'testing' facility and not a commercial industrial site, the lessons learned are important to consider. Current research in Oregon is focused on the collaborative OE permitting processes.

Key findings from this research mirrored lessons learned from previous research on consultation:[50]

- **The design and timing of the consultation is important.** Well planned, early and frequent engagement to keep stakeholders and the public informed and involved is key. Equally important are using appropriate methods for information sharing *and* listening to perceptions and interests. Provide meaningful responses to stakeholder input.
- **Be 'community driven' versus 'developer driven'.** Part of this is knowing *whom to engage and why*. While public relations focuses on 'the public', engagement focuses on the community and the stakeholders there (individuals and organizations). Basic rule of thumb: be inclusive. Stakeholders need understandable information so that they can provide input and form opinions. *Trust* is another part. Trust has been defined as "the willingness to rely on those who have the responsibility for making decisions and taking actions".[51] Trust is an important component of success and can reduce the amount of active opposition to a project. Partnering with a local organization to learn about the community and become familiar with priority issues facing the community might also be helpful to build trust, assist in the information sharing needed to plan and site and lessen impact and/or increase the benefits to stakeholders and others.
- **Engagement requires adequate capacity, resources and time.** Each community is unique, yet the three foundational principles for this are to (1) budget for consultation, (2) have staff and/or utilize local people who are skilled in working with communities to build trust/rapport upfront, (3) become familiar with priority issues facing the community as early as possible.

7.2.1 Conclusion

Research that studies and documents lessons learned about connection between people (users), between people and place (ocean), and between economic opportunities and place (testing facilities versus commercial industrial sites) is filled with strategies that can be considered and modelled. Putting into practice lessons learned from research regarding consultation is not only a good idea, it can be instrumental in the design and implementation of future consultation processes that reach their desired outcomes. This requires a good design, a lot of upfront work and periodic evaluation and adjustment as needed. Viewing the community as partners who provide local knowledge and different ways of understanding is important to plan for and move forward with OE projects.

7.3 Working towards a social license for tidal energy development in Nova Scotia's Bay of Fundy

The Bay of Fundy, bordered by Nova Scotia, New Brunswick and Maine, is home to the world's highest tidal range averaging 16 m. The tidal flow in the Bay of Fundy is the collective equivalent of the world's major rivers, and where in-stream tidal turbines are being tested in the New Minas Passage, current speeds are in excess of 5 m/s. Significant amounts of energy can be harnessed from these tides and currents,[52] and, given this potential, the Fundy Ocean Research Centre (FORCE) was developed to serve as a test centre for the in-stream tidal turbines in the New Minas Passage. While the FORCE site represents large-scale tidal energy development opportunities, southwest Nova Scotia is also home to small-scale tidal energy development exploration. Both small- and large-scale tidal energy development is being supported by feed-in tariffs. This case study explores the role of consultation and engagement in the development of strategic planning documents and policies that provided guidance to tidal energy development in the Bay of Fundy. More specifically, the case study highlights the role of consultation in supporting a social license for tidal energy development in Nova Scotia.

Extracting energy from the powerful tides in places like the Bay of Fundy have been explored for over a century. But it's only been in the last several decades where development has occurred. The Annapolis River Tidal Generating Station demonstration project came online in 1984, one of only two grid-connected tidal power plants built at the time. Significant insights have emerged from research exploring the consequences of barrage-style developments like the Annapolis River Tidal Generating Station. Current tidal energy development in Nova Scotia's Bay of Fundy focuses on tidal stream (or hydrokinetic) power. Potential benefits to the Province of Nova Scotia from in-stream tidal energy development include:

- Meeting its renewable energy targets of 40% by 2020;
- Energy security and long-term price stability;
- Reduction of environmental impacts from other forms of energy production;
- Opportunities to enhance and to develop supply chain with respect to tidal energy industry;
- Opportunities to support broader sustainable economic development goals locally and regionally; and
- The ability not only to build generating capacity in an incremental way that provides opportunities for early revenue generation but also to employ adaptive management techniques.

Nova Scotia's vision for tidal energy development has been guided, in part, by the Nova Scotia Marine Renewable Energy Strategy (2012) and an earlier Department of Energy discussion paper on Marine Renewable Energy Legislation for Nova Scotia (2010). These documents discuss the regulatory context for tidal energy development and how this type of development should proceed. Preceding this work was the development of a strategic environmental assessment (SEA) for Tidal Energy Development in Nova Scotia and an SEA for Tidal Energy Development Update.[53] SEAs are viewed as a mechanism to shift the regulatory approach from policymaking to policy application.[54] The SEA is:

> intended to ensure positive contributions to sustainability, as well as mitigation of adverse environmental effects; to enhance the openness and credibility of strategic level decision-making; to provide earlier, clearer and more reliable guidance for the planning and approval of particular projects and other subsequent undertakings; and to improve the overall efficiency and fairness as well as the effective quality of decision-making.[55]

The Province of Nova Scotia used the SEA process to guide the development of the regulatory process with respect to tidal energy development.[56] In addition to developing a state-of-knowledge understanding of the environmental and socio-economic impacts associated with tidal energy development based on an exhaustive and comprehensive review of literature, the SEA process was supported by extensive community consultation in six communities near to or directly adjacent to the proposed tidal energy development sites. Additional consultation sessions were held with key stakeholders, including First Nation groups.

The outcomes of the SEA process resulted in 29 recommendations to guide the development of tidal energy in Nova Scotia. Examples of recommendations include the adaptation of sustainability principles, a Mi'kmaq Ecological Knowledge Study (MEKS), an incremental development approach, public education and awareness, relevant legislation to guide tidal energy development, and community participation and benefits.[57] These recommendations were revisited in 2013 with a strategic environmental assessment (SEA) for Tidal Energy Development – Bay of Fundy Update. Of the 29 recommendations noted in the earlier SEA, most have been addressed, in particular the need to approach tidal energy development incrementally, the use of an adaptive management approach and consultation with First Nation communities.

The Mik'maq ecological studies were carried out in two phases exploring both the outer and inner regions of the Bay of Fundy. These studies documented the cultural significance of the Bay of Fundy to the Mi'kmaq and also of traditional and contemporary regions for the harvesting of marine life such as mackerel, clams and lobster. The SEA 2013 update consulted specifically with the Confederacy of Mainland Mi'kmaq's Conservation Group. A lack of trust was noted as a key concern, coupled with frustration expressed over the lack of long-term and ongoing consultation with the Mi'kmaq. Developing baseline research related to fish interactions with tidal turbines and associated behaviour emerged as a key concern as well.

Two other key documents, supported by extensive community consultation, were developed. In the *Community & Business Toolkit for Tidal Energy Development*,[58] the 13 modules comprising the toolkit were developed, in part, by direct engagement with community, industry and government stakeholders over key issues in tidal energy development in Nova Scotia. An issue emerging from this process was the lack of support for communities engaged in small-scale tidal energy development related to engagement. The follow-up *Tidal Energy Community Engagement Handbook* was developed to address this concern.[59] The development of the handbook, directly supported by three communities involved in small-scale tidal energy development, was specifically used as an opportunity to develop engagement strategies for these communities.

While the SEA, MEKS, the toolkit, and engagement handbook provided significant insight into stakeholder concerns about how tidal energy development should proceed, lacking was insight into the potential economic impact associated with tidal energy development – a key issue identified by stakeholders. Understanding this potential is significant for Nova Scotia. In the recent report, *Now or Never: An Urgent Call to Action for Nova Scotians*,[60] Nova Scotia was noted as being in a period of significant population loss and economic decline, especially in its rural regions. The report highlighted the importance of significant and immediate action, hence the title of the report. The *Value Proposition for Tidal Energy*, commissioned by the Nova Scotia Offshore Energy Research Association, explored the business case for tidal energy that identified the potential value, broader economic benefits and costs, and economic impacts of tidal power through a range of three scenarios:[61] low (67 MW), moderate (300 MW) and high (500 MW). If Nova Scotia were able to reach the moderate goal of 300 MW, this would provide significant economic impacts. In this scenario, tidal energy development would contribute over $1.6 billion to the Nova Scotia's GDP, create 21,933 full-time equivalent (FTE) jobs and generate $1.1 billion in labour income. Included in these figures is an overall investment of $500 million to $2.5 billion in tidal energy infrastructure from 2015 to 2040.

Tidal energy development in Nova Scotia has largely been guided by consultation with stakeholder groups in the development of the SEA, MEKS, and other critical documents. This work, in part, has supported the legitimacy of tidal energy development in Nova Scotia as government and industry have followed through on commitments made to its stakeholders. Legitimacy is considered a "condition reflecting cultural alignment, normative support, or consonance with relevant rules or laws"[62] and as "the process by which key stakeholders, the general public, key opinion leaders, or government officials accept a venture as appropriate and right".[63] Developing a social license begins with establishing the legitimacy of a project, which implies acceptance but not necessarily approval.

Establishing credibility and trust are prerequisites for broader approval of tidal energy development in Nova Scotia, and this will be challenging given that trust in government and in industry is low.

While Nova Scotia has steadily worked towards developing a social license for tidal energy development and has to a fair degree established the legitimacy of this type of development, it has yet to gain the credibility and trust necessary for acceptance of this type of energy development. Recent events have set back broader support and approval of tidal energy development in Nova Scotia. Members of the Nova Scotia Inland Fisheries Association, as well as other commercial fishing organizations, have filed an injunction to halt the deployment of the latest tidal energy turbine test in New Minas Passage, citing a lack of consultation and inadequate baseline research regarding fish behaviour and distribution, as well as a lack of knowledge related to fish and turbine interactions. While these groups were included in earlier consultation processes, as the test centre grew (from 2 MW up to 300 MW), there was a noticeable gap in consultation, according to key players in the commercial fishing community, and it was felt that industry and government lacked the social license to proceed with tidal energy development. The term 'social license' was evoked by the fishing community in this instance. Despite the best intensions to live up to its promises of ongoing and inclusive stakeholder consultation, government and industry fell short of this promise, according to many people in the commercial fishery, and the fallout has been the delay of deployment of the test turbine.

Key lessons can be drawn from this brief review of tidal energy development in Nova Scotia:

- The role of strategic environmental assessments is critical in developing a broad and comprehensive knowledge base necessary for the development of regulations, policies and procedures for the development of tidal energy. SEAs that are framed within a stakeholder consultation framework are particularly important for developing insights into key concerns and issues with respect to development.
- Strategic environmental assessments and other consultation processes that might support other studies (e.g. Mi'kmaq ecological studies, toolkit, etc.) do not necessarily equate to ongoing, broad-based consultation and public education regarding a development project. Rather, these consultation processes are in reference to the development of a specific report or document.
- Broad-based, inclusive and ongoing consultation and public education are critical in supporting a social license for tidal energy development in Nova Scotia. Public education, specifically, is critical in developing greater energy literacy among the broader public.
- Despite the ongoing concerns related to the vagueness of the concept of social license, at the very least the term allows industry, government, communities and the larger public space to articulate the importance of the social contract that should be forged among industry, a community, government and the larger public.

8 Conclusion: lessons for ocean energy consultation

Emerging from the case studies in Guernsey, Oregon and Nova Scotia is the critical importance of understanding peoples' connection to place. While the case studies highlighted a range of other important consultation issues, the issue of place connectedness and the need to understand its influence on perspectives on technology, the location of energy development, the scale of development, the value of the sea for traditional and spiritual sustenance and the nature of impacts is fundamental. By working to better understand these place-connected issues more meaningfully and with perhaps more creative and meaningful methods of consultation, proponents of OE development may better lay the foundations of trust and respect.

Highlighted are several lessons that are important to convey for moving towards greater social acceptance of OE development:

- It is important that developers cease to view consultation merely as a legal obligation. This approach is limiting, and treating the public as a hostile threat to project completion may well become a self-fulfilling prophecy.[64] A great deal of evidence suggests that consultation, done properly, can improve project outcomes, lead to greater acceptance (for instance by finding socially acceptable sites, as in the Guernsey case study) and address issues in planning that the developer would otherwise be unaware of – and liable for. Thus, changing how proponents view consultation is necessary if the positive outcomes of consultation are the desired ends. Using visual, participatory methods like the ones in the Guernsey case study can play a key enabling role in facilitating constructive and positive consultation.
- Participatory consultation processes must ensure that social learning about energy issues is incorporated into a broader and more comprehensive manner than it has been to date. The extent to which this is true may vary with jurisdiction, as energy literacy is unlikely to be uniformly distributed in a population.[65] What is required, then, is an adaptive approach to consultation, mixing information provision, social learning and opportunities for feedback and involvement as appropriate to the public audience being engaged. The strategic environmental assessment (SEA) process in Nova Scotia worked to build energy literacy with respect to tidal energy development in Nova Scotia and provided opportunities for feedback and involvement. Overall, the process reflected an adaptive approach in that the methods of consultation, the stakeholders consulted and the methods for receiving feedback were flexible, open to change, and modified during the process.
- Without trust between communities and developers, OE projects are unlikely to be completed. To develop this trust, several things must be done. Contact between the proponents and the host community for a project should happen as early as possible.[66] Several instances where a community lacked early contact or local input on planning until after a project had been given

governmental approval show that such decisions lead to vocal opposition.[67] However, developers may still be attracted to focusing on deals with higher governmental levels, in part due to beliefs that the public will not be willing to trust private interests.[68] It is important that such beliefs be addressed, as trust between developers and communities has been shown by these three case studies to be possible, but it cannot come about without engagement.

- One important aspect of ocean energy consultation evident in these case studies is that it often takes place at an early stage – due to the current stage of development of OE technology (which means that projects are often pilots or evolve as they are being proposed). *This contrasts with established consultation practices on onshore wind energy projects, which are perhaps more routine and often take place at a relatively late stage of project development.* This kind of consultation brings additional challenges, as pilot projects may evolve into much larger local energy projects (as seen in Nova Scotia), or public views are sought on an array of hypothetical projects to inform future policy and priorities (Guernsey), or they are sought at a stage when it is not even clear what the technology will be exactly (Oregon). Nevertheless, even in such a challenging context, these case studies have shown that a degree of trust can be gained (Oregon and Nova Scotia), and public input can be obtained that may effectively lead to more acceptable projects (Guernsey). This means that even if OE projects may not be entirely clear or finalized, it is still important to begin consultation at an early stage of the development process.

The emergence of OE development is timely. Firstly, OE development has significant potential for addressing world energy needs, reducing the impacts of climate change, supporting economic development, and serving as a platform for the development of new industries and technologies. Secondly, its emergence parallels the growing issue and importance of empowerment of communities and the public over development issues that affect their lives. Consultation and its related principles, strategies, and tools serve as a bridge by which communities and the public can engage in discourse related to OE development in their communities.

Notes

1 Ricci, M., Bellaby, P., and Flynn, R. (2010) Engaging the Public on Paths to Sustainable Energy: Who Has to Trust Whom? *Energy Policy*, 38, pp. 2633–2640; Wiersma, B., and Devine-Wright, P. (2014) Public Engagement with Offshore Renewable Energy: A Critical Review. *WIREs Climate Change*, 5(4), pp. 493–507.
2 Walker, G., et al. (2011) Symmetries, Expectations, Dynamics and Contexts: A Framework for Understanding Public Engagement with Renewable Energy Projects. In: Devine-Wright, P. (ed.) *Renewable Energy and the Public: From NIMBY to Participation*. London: Earthscan, pp. 1–14.
3 Walker, G., et al. (2011) n. 2.
4 Ricci, M., Bellaby, P., and Flynn, R. (2010) n. 1; Adams, M., Wheeler, D., and Woolston, G. (2011) A Participatory Approach to Sustainable Energy Strategy Development in a Carbon-intensive Jurisdiction: The Case of Nova Scotia. *Energy Policy*, 39, pp. 2550–2559.

5 Adams, M., Wheeler, D., and Woolston, G. (2011) n. 4.
6 West, J., Bailey, I., and Winter, M. (2010) Renewable Energy Policy and Public Perceptions of Renewable Energy: A Cultural Theory Approach. *Energy Policy*, 38, pp. 5739–5748.
7 Stirling, A. (2008) "Opening up" and "Closing down" Power, Participation, and Pluralism in the Social Appraisal of Technology. *Science, Technology & Human Values*, 33(2), pp. 262–294; Glucker, A. N., et al. (2013) Public Participation in Environmental Impact Assessment: Why, Who and How? *Environmental Impact Assessment Review*, 43, pp. 104–111.
8 Glucker, A. N., et al. (2013) n. 7.
9 Ibid.
10 Ibid.
11 Ibid.
12 Ricci, M., Bellaby, P., and Flynn, R. (2010) n. 1.
13 Barnett, J., et al. (2012) Imagined Publics and Engagement Around Renewable Energy Technologies in the UK. *Public Understanding of Science*, 21(1), pp. 36–50.
14 Ricci, M., Bellaby, P., and Flynn, R. (2010) n. 1.
15 Devine-Wright, P. (2009) Rethinking NIMBYism: The Role of Place Attachment and Place Identity in Explaining Place-Protective Action. *Journal of Community and Applied Social Psychology*, 19, pp. 426–441; Wiersma, B. (2016) *Public Acceptability of Offshore Renewable Energy in Guernsey: Using Visual Methods to Investigate Local Energy Deliberations*. Unpublished thesis, Exeter University. Available at: https://ore.exeter.ac.uk/repository/handle/10871/21565
16 Delvaux, P. A. G., Rabuteau, Y., and Stanley, K. (2013) *Civil Society Involvement and Social Acceptability of Marine Energy Projects: Best Practices of the Marine Energy Sector*. France: MERiFIC.
17 Batel, S., Devine-Wright, P., and Tangeland, T. (2013) Social Acceptance of Low Carbon Energy and Associated Infrastructures: A Critical Discussion. *Energy Policy*, 58, pp. 1–5; Delvaux, P. A. G., Rabuteau, Y., and Stanley, K. (2013) ibid.
18 Delvaux, P. A. G., Rabuteau, Y., and Stanley, K. (2013) n. 16; Wiersma, B., and Devine-Wright, P. (2014) n. 1.
19 del Rio, P., and Burgillo, M. (2009) An Empirical Analysis of Renewable Energy Deployment on Local Sustainability. *Renewable and Sustainable Energy Reviews*, 13, pp. 1314–1325.
20 Moffat, K., and Zhang, A. (2014) The Paths to Social License to Operate: An Integrative Model Explaining Community Acceptance of Mining. *Resources Policy*, 39, pp. 61–70.
21 Thomson, I., and Joyce, S. (2008) The Social License to Operate: What It Is and Why Does It Seem So Difficult to Obtain? In: *PDAC Convention*. Toronto. Available at: http://oncommonground.ca/wp-content/downloads/PDAC_2008_Social_Licence.pdf
22 Parsons, R., Lacey, J., and Moffat, K. (2014) Maintaining Legitimacy of a Contested Practice: How the Minerals Industry Understands Its 'Social Licence to Operate'. *Resources Policy*, 41, pp. 83–90.
23 Assefa, G., and Frostell, B. (2007) Social Sustainability and Social Acceptance in Technology Assessment: A Case Study of Energy Technologies. *Technology in Society*, 29(1), pp. 63–78.; Bidwell, D. (2013) The Role of Values in Public Beliefs and Attitudes Towards Commercial Wind Energy. *Energy Policy*, 58, pp. 189–199; Batel, S., Devine-Wright, P., and Tangeland, T. (2013) n. 17.
24 Hall, V., et al. (2013) Early Embryonic Development, Assisted Reproductive Technologies, and Pluripotent Stem Cell Biology in Domestic Mammals. *The Veterinary Journal*, 197(2), pp. 128–142; Walker, G., and Devine-Wright, P. (2008) Community Renewable Energy: What Should It Mean? *Energy Policy*, 36(2), pp. 497–500; Venkatesh, V., et al. (2003) User Acceptance of Information Technology: Toward a Unified View. *MIS Q*, 27, pp. 425–478.

25 Vorkinn, M., and Riese, H. (2001) Environmental Concern in a Local Context: The Significance of Place Attachment. *Environmental Behaviour*, 33, pp. 249–263.
26 Douvere, F. (2008) The Importance of Marine Spatial Planning in Advancing Ecosystem-Based Sea Use Management. *Marine Policy*, 32(5), pp. 762–771.
27 Pomeroy, C., Hall-Arber, M., and Conway, F. (2015) Power and Perspective: Fisheries and the Ocean Commons Beset by Demands of Development. *Marine Policy*, 61, pp. 339–346; Sullivan, C. M., et al. (2015) Combining Geographic Information Systems and Ethnography to Better Understand and Plan Ocean Space Use. *Applied Geography*, 59, pp. 70–77.
28 Wiersma, B. (2016) n. 15.
29 Wiersma, B., and Devine-Wright, P. (2014) n. 1.
30 Devine-Wright, P. (2011a) *Renewable Energy and the Public*. London: Earthscan.
31 Devine-Wright, P., and Howes, Y. (2010) Disruption to Place Attachment and the Protection of Restorative Environments: A Wind Energy Case Study. *Journal of Environmental Psychology*, 30, pp. 271–280; Firestone, J., and Kempton, W. (2007) Public Opinion About Large Offshore Wind Power: Underlying Factors. *Energy Policy*, 35(3), pp. 1584–1598.
32 Devine-Wright, P. (2011a) n. 30.
33 Devine-Wright, P. (2009) n. 15; Batel, S., and Devine-Wright, P. (2015) A Critical and Empirical Analysis of the National-Local 'Gap' in Public Responses to Large-Scale Energy Infrastructures. *Journal of Environmental Planning and Management*, 58(6), pp. 1076–1095.
34 Devine-Wright, P., and Howes, Y. (2010) n. 31.
35 McLachlan, C. (2009) 'You Don't Do a Chemistry Experiment in Your Best China': Symbolic Interpretations of Place and Technology in a Wave Energy Case. *Energy Policy*, 37, pp. 5342–5350.
36 See Wiersma, B. (2016) n. 15.
37 Johnsen, S., May, J., and Cloke, P. (2008) Imag(in)ing 'Homeless Places': Using Auto-Photography to (Re)examine the Geographies of Homelessness. *Area*, 40(2), pp. 194–207.
38 Wiersma, B. (2016) n. 15.
39 Walker, G., et al. (2010) Renewable Energy and Sociotechnical Change: Imagined Subjectivities of 'the Public' and Their Implications. *Environment and Planning A*, 42, pp. 931–947.
40 McLachlan, C. (2009) n. 35.
41 Hagerman, G., and Bedard, R. (2003) *Guidelines for Preliminary Estimation of Power Production by Offshore Wave Energy Conversion Devices*. EPRI WP-001. Palo Alto, CA: EPRI.
42 Hagerman, G., et al. (2004) *E21 EPRI Survey and Characterization of Potential Offshore Wave Energy Sites in Oregon*. WP-OR-003. Palo Alto, CA: EPRI.
43 Ibid.
44 Yin, Y. (2009) *Is Wave Energy Comparatively Sustainable in Oregon?* Thesis, Oregon State University. Available at: https://ir.library.oregonstate.edu/xmlui/bitstream/handle/1957/12116/yaoyin.pdf
45 Brekken, T. (2009) *Wave Energy Opportunities and Developments* PowerPoint Presentation Slides, retrieved from Oregon State University Wallace Energy Systems and Renewables Facility. Available at: http://eecs.oregonstate.edu/wesrf/
46 Henkel, S. K., et al. (2013) Environmental and Human Dimensions of Ocean Renewable Energy Development. *Proceedings of Institute of Electrical & Electronics Engineers*, 101(4), pp. 991–998.
47 Conway, F., et al. (2009) *Socio-Economic Perspectives of Wave Energy Development: Final Report to the Oregon Wave Energy Trust*. Available at: http://oregonwave.org/oceanic/

wp-content/uploads/2013/05/Socio-Economic-Perspectives-of-Wave-Energy-2009. pdf; Conway, F., et al. (2010) Ocean Space, Ocean Place: The Human Dimensions of Wave Energy in Oregon. *Oceanography*, 23(2), pp. 82–91.

48 Industrial Economics, Inc. (2012) *Identification of OCS Renewable Energy Space-Use Conflicts and Analysis of Potential Mitigation Measures*. Herndon, VA: U.S. Department of the Interior, Bureau of Ocean Energy Management; Sullivan, B. C. M., et al. (2015) n. 27; Pomeroy, C., Hall-Arber, M. and Conway, F. (2015) n. 27.

49 Goodwin, B. (2015) *Evaluating Community Engagement in Wave Energy Siting Off the Oregon Coast*. Unpublished master's thesis, Oregon State University. Available at: http://ir.library.oregonstate.edu/xmlui/handle/1957/56347

50 Blahna, D. J., and Yonts-Shepard, S. (1989) Public Involvement in Resource Planning: Toward Bridging the Gap Between Policy and Implementation. *Society & Natural Resources*, 2(1), pp. 209–227; Chase, L. C., et al. (2004) Public Participation in Wildlife Management: What Do Stakeholders Want? *Society & Natural Resources*, 17(7), pp. 629–639; Richards, C., Carter, C., and Sherlock, K. (2004) *Practical Approaches to Participation*. Aberdeen: Macaulay Institute; Senecah, S. (2004) The Trinity of Voice: The Role of Practical Theory in Planning and Evaluating the Effectiveness of Environmental Participatory Processes. In: Depoe, S. P. (ed.) *Communication and Public Participation in Environmental Decision Making*. Albany: State University of New York Press, pp. 13–34; Ehler, C., and Douvere, F. (2009) *Marine Spatial Planning: A Step-by-Step Approach Toward Ecosystem-Based Management*. UNESCO; Chozas, J. F., Stefanovich, M. A., and Sørensen, H. C. (2010) Toward Best Practices for Public Acceptability in Wave Energy: Whom, When and How to Address. *Group*, 7, pp. 1–8; Olsen, C. S., and Shindler, B. A. (2010) Trust, Acceptance, and Citizen–Agency Interactions After Large Fires: Influences on Planning Processes. *International Journal of Wildland Fire*, 19(1), pp. 137–147; Johnson, A., et al. (2013) Heterochromatic Gene Silencing by Activator Interference and a Transcription Elongation Barrier. *Journal of Biological Chemistry*, 288(40), pp. 28771–28782.

51 Siegrist, M., and Cvetkovich, G. (2000) Perception of Hazards: The Role of Social Trust and Knowledge. *Risk Analysis: An International Journal*, 20(5), pp. 713–720.

52 Karsten, R. H., et al. (2008) Assessment of Tidal Current Energy in the Minas Passage, Bay of Fundy. *Proceedings of the Institution of Mechanical Engineers: Part A: Journal of Power and Energy*, 5(222), pp. 493–507.

53 See Doelle, M. (2009) The Role of Strategic Environmental Assessments (SEAs) in Energy Governance: A Case Study of Tidal Energy in Nova Scotia. *Journal of Energy & Natural Resources Law*, pp. 111–144.

54 See Partidário, M. R. (2003) *Strategic Environmental Assessment (SEA): Current Practices, Future Demands and Capacity-building Needs*. Lisbon: International Association for Impact Assessment.

55 Benevides, H. et al. (2009) Law and Policy Options for Strategic Environmental Assessment in Canada. *SSRN Electronic Journal*. DOI: 10.2139/ssrn.1660403. Available at: https://www.researchgate.net/publication/228203808_Law_and_Policy_Options_ for_Strategic_Environmental_Assessment_in_Canada

56 Oldrieve, M. E. (2013) The Role of Strategic Environmental Assessments for Emerging Marine Renewable Energy Sectors: The Nova Scotian Example. *Journal of Environmental Assessment Policy & Management*, 15(2), pp. 1–25.

57 Colton, J., et al. (2016) *Energy Projects, Social License, Public Acceptance and Regulatory Systems in Canada: A White Paper*. The School of Public Policy SPP Papers, University of Calgary.

58 Colton, J., and MacDougal, S. (2013) *The Community and Business Toolkit for Tidal Energy Development*. Wolfville, Nova Scotia: Acadia Tidal Energy Institute.

59 Isaacman, L., and Colton, J. (2013) *Tidal Energy Community Engagement Handbook.* Wolfville, Nova Scotia: Acadia Tidal Energy Institute.
60 Ivany, R., et al. (2014) *The Report of the Nova Scotia Commission on Building Our New Economy.* Nova Scotia: Nova Scotia Commission on Building Our New Economy. Available at: https://cra.ca/wp-content/uploads/2014/02/Commission-on-Building-a-New-Economy-.pdf
61 Gardner Pinfold Consultants Inc., Acadia Tidal Energy Institute and Offshore Energy Research Association (2015) *Value Proposition for Tidal Energy Development in Nova Scotia, Atlantic Canada and Canada.* Halifax: Authors.
62 Scott, W. R. (1995) *Institutions and Organizations.* Thousand Oaks, CA: Sage, p. 45.
63 Aldrich, H., and Fiol, C. M. (1994) Fools Rush in? The Institutional Context of Industry Creation. *Academy of Management Review,* 19(4), pp. 645–670.
64 Fast, S., and Mabee, W. (2015) Place-making and Trust-building: The Influence of Policy on Host Community Responses to Wind Farms. *Energy Policy,* 81, pp. 27–37; Barnett, J., Lambert, S., and Fry, I. (2017) The Hazards of Indicators: Insights from the Environmental Vulnerability Index. *Annals of the Association of American Geographers,* 98(1), pp. 102–119.
65 Student Energy (2015) *Global Energy Literacy Research Project.* Available at: www.studentenergy.org/images/downloads/SE-GELRP.pdf
66 Delvaux, P. A. G., Rabuteau, Y., and Stanley, K. (2013) n. 16.
67 Hindmarsh, R., and Matthews, C. (2008) Deliberative Speak at the Turbine Face: Community Engagement, Wind Farms, and Renewable Energy Transitions in Australia. *Journal of Environmental Policy and Planning,* 10(3), pp. 217–232; Fast, S., and Mabee, W. (2015) n. 70.
68 Barnett, J., Lambert, S., and Fry, I. (2017) n. 70.

Chapter 13

Ocean energy at the edge

Laura Watts and Brit Ross Winthereik

How do we change perceptions
of 'the edge'
and marine energy.

There is needed
a quite new attitude
about (marine) energy.

Is it surprising to people from Orkney
that Danish people (or anyone else)
are unaware of the value of marine energy –

What you know in daily experience,
that makes you understand degrees
of importance. I suggest that by linking
the two places, and providing an 'edge'
for Denmark (as was suggested) and
a European centre for Orkney
both places will benefit.

I Introduction

The preceding inscriptions on Post-it® notes were written by marine energy industry representatives from Denmark and Orkney, Scotland, to Marine Renewable
Energy at Futures Edge: Connecting Denmark and Orkney, Scotland, a workshop
held in Copenhagen in September 2011. This group of industry, policy and academic practitioners were brought together in an interdisciplinary workshop to
explore what the *futures' edge* might be for marine energy in these two places.

The word 'edge' in the title for the workshop was chosen with deliberate care.
We both work as ethnographers of ocean energy (OE), and as such we hear 'edge'

over and over. The word permeates conversations, presentations and documentation in our ethnographic collaborations with people, places and organizations. For example, marine energy is often talked about as though positioned at the edge of a larger energy industry, with fossil fuel power stations at the centre. Secondly, marine energy is an environmental resource located at the geographic periphery. Those who work in the industry live, or at least must work, where these high-energy seas are to be found: literally at the geographic edge. In both these examples, marine energy is conceptualised as marginal compared to some other central power.

The workshop in 2011 was the beginning of a research project to explore this edgy quality that suffuses marine renewable energy, their people and places. Since then, we have continued the exploration as part of a research project entitled Marine Renewable Energy as Alien: Social Studies of an Emerging Industry. We have spent time with people in Denmark and in Orkney and have talked with them about what it means to work in an energy sector that is both 'in the making' and marginalised.

This chapter will take up the call made by participants in the workshop and do two things. First, we will explore the many different versions of the edge in ocean energy we have encountered during our four-year research project. Second, we will draw on resources in the social studies of science and technology and the ethnographic studies of infrastructure in order to reconceptualise or 'reconfigure'[1] the ocean energy edge not as marginal but as powerful and with tactical advantages that may be relevant to other ocean energy edges and communities throughout the world.

2 Ethnography at/of the edge

Over the next four years after the workshop, we spent considerable time in conversation, sometimes immersed, in the organizations and places around wave and tide energy in Denmark and in Orkney, Scotland. We gathered extensive transcripts and notes from interviews and meetings we participated in, took thousands of photographs, read and annotated hundreds of related documents. As ethnographers, we attended to the social and material infrastructure of wave and tide energy.[2] Brit Ross Winthereik worked with people around the Danish wave energy industry; Laura Watts worked in Orkney with those working in and around the marine energy industry there.

We were interested, first, in how these new energy infrastructures were developing in social, cultural, technical and environmental relations – in policy and practice, for example. In this, we draw on work in science and technology studies (STS) that understands infrastructure as not just a material system of transportation, nor just as a technical object, but also as embedding social, cultural and political decisions and agencies that have to hold together and endure over long periods of time.[3] In the context of marine energy, attending to infrastructures as social and technical meant attending to how, in our fieldsites, political decisions (whether local, industry or national politics) hindered or enabled OE to be transformed and transmitted onto the electricity grid. Second, we were exploring how local communities, which often included the marine energy industry community itself, were developing a relationship to these 'alien' and 'other' infrastructures for marine energy in their seascape – from the devices to the onshore buildings – and, conversely, how the devices themselves moved through different environments making relations with business investors, politicians, actual and potential subcontractors, and industry partners. In short, how was tide and wave energy being enacted at the edge, and how was it being done differently in Orkney, Scotland, compared to Denmark?

3 Edge histories

Orkney and Denmark already had an existing relationship prior to our research, of course, and interestingly each has a very different role in the overall global marine energy industry.

Orkney, the islands off the far northeast coast of Scotland, is a place of very high energy waves and strong tides; waves may be 14 m high and the tide perhaps 8 knots. This is the location for the European Marine Energy Centre (EMEC), the world's first and longest running grid-connected test site for wave and tide energy generators. It has had large and full-scale prototypes in the water and on the grid since 2004, when the Pelamis 'sea snake' became the first wave energy machine in the world to generate on-grid electricity. Since then, EMEC has expanded to provide 14 berths for wave and tide energy generators to plug in and test in the open sea. At our last count, in 2012, there have been 45 device developers testing at EMEC, with well known energy and engineering companies involved, from Kawasaki to Voith, from Alstom to E.ON. Around 300 people work in or around the test site in Orkney, an ecosystem of support companies and facilities that is almost unique in the industry. Although there are perhaps 40 open sea marine energy test sites around the world, none have comparable experience; the number of device developers testing across all of the United States test sites put together is under 20, at the time of writing. This is why EMEC and Orkney are regarded by many as the industry beacon. The European Marine Energy *Centre* is often taken as just that, the global *centre* for the marine energy industry.

In comparison, the waves surrounding Denmark are much smaller than those near Orkney. Still, in 1998 the combination of more than 8,000 km of coastline, a history of renewable energy production and a group of inventive marine energy enthusiasts led to the design of the Danish Wave Energy Program. In its first years, the programme allocated small amounts of state funding to around 50 marine energy projects. Roughly ten of these received additional funding to conduct further development and testing. Over the years, testing activity flourished, creating a need to further expand the testing infrastructure. At first, this infrastructure was set up by a group of committed wave energy inventors (The Wave Energy Association). This was a grassroots movement, and its main objective was to select and lay down cabling for a test site. In 2008, two new state-led funding programmes were initiated, and a handful of new devices were designed, along with an increasing need to perform offshore tests. In 2009, the company Wavestar began testing its 1:2-scale device in the North Sea just outside the port of Hanstholm. This led to considerable publicity for the emerging wave energy sector. A year later, in 2009, as a result of collaboration between the port of Hanstholm, the local municipality, the local business association and Aalborg University, the Danish Wave Energy Center (DanWEC) was officially announced right where the Wavestar was also based. In 2012, a national wave energy strategy was published, but in 2016 Wavestar went bankrupt. Several other companies are testing in the North Sea and in other places around Denmark, but none of them are grid connected. With DanWEC not managing to establish itself as *the* centre for marine energy testing in Denmark, and with Wavestar gone, there is no one place where Danish marine energy is happening but many.

In 2011, the Danish marine energy inventors explicitly looked to EMEC and Orkney as their centre, as our Post-it notes showed. And with the situation today, it would appear that the relationship between Danish and Scottish marine renewable energy industries has EMEC and Orkney at the centre, with DanWEC and Denmark located more at the periphery.

However, this comparison belies some crucial histories that relocate this relationship and the histories of renewable energy more generally. On 9 July 2015, Denmark generated over 140% of its electricity from wind energy. This was a day when the entire country ran on nothing but air and also generated considerable revenue by exporting its excess wind energy to Norway, Sweden and Germany. Denmark could therefore be said to be synonymous with wind energy: it has some of the world's largest wind energy companies (e.g. Vestas) and massive national support for the industry. Wind energy is celebrated and entangled in Danish pride for their world-leading energy technology industry. The Danish Government has invested in and remained committed to the development of this renewable energy industry since the 1980s. Wind energy dominates the energy landscape, both in policy and practice, with the Danish Wind Energy Association, a partnership of over 240 companies, providing considerable political and commercial leverage. In Denmark, wind energy is the energy source at the *centre*, with wave, solar and biomass more in the *periphery*. Compared to solar and biomass, marine energy is marginal, but it is at least 'on the map' of this renewable energy-led national policy.

Back in Orkney and in the UK more generally, the political geography for renewable energy is somewhat muddier. Orkney was the location for the UK's largest experimental wind turbine in the 1980s (300 kW), but the Government failed to commit to wind energy and research stalled. Although tariffing schemes that privilege renewable energy have been in place, such as the Renewable Obligation Certificates (ROCs), it could not be said that wind energy or any other renewable energy lies at the centre of UK energy policy. Marine energy is not just marginal to wind energy, the entire renewable energy industry is somewhat peripheral to an energy policy that favours centralised power sources (nuclear, gas and 'clean' coal) and is moving towards importing electricity and interconnector grid networks.[4]

Comparing these histories and political geographies of renewable energy, then, suggests that Denmark, with its long heritage of wind energy, lies very much at the global centre of renewable energy, with the UK a poor peripheral figure. It is into this wider political geography, which has differing national positions, that the marine energy industry must locate itself. How centres and peripheries are constructed and can be constructed differently in the marine energy industry are what we will now explore.

4 Centre–periphery

There is an extensive literature across sociology, anthropology and geography that reflects on how the centre–periphery dichotomy is made and how it might be unmade, or at least represented and read otherwise. These largely derive from discussions of marginalisation, how places and peoples are made marginal to some central power through colonisation or other political histories of 'othering' – which makes both 'us' and 'other'.[5] Two particularly important threads in that discussion, which seem important for centre–periphery in marine energy, are (1) the effects of physical geography and its mapping; and (2) the discourse of urban nodes versus rural edges perpetuated in notions of globalisation and models like the Network Society.[6]

First, in the mentioned literature maps are technologies that represent space but also politics. When the Flemish geographer Mercator presented his map of the world in the sixteenth century, his projection was intended to aid seafarers' voyages across the Atlantic by representing lines of constant course. The prime meridian line runs straight through London and down through Spain and Algeria on this map, placing the European colonial powers at the centre. But by supporting colonial sea navigation, landforms are completely misrepresented: Northern Europe is massively over-proportioned compared to the African continent. It makes Europe and North America look much more dominant on the planet than they actually are. It is this outdated and misleading map of the world that remains more or less the standard, since its politics of centre–periphery privileges current political world powers.[7] Moreover, such politics of maps reproducing a central power happens at the national scale. Maps of the UK are centred on London as the capital, and the choice of viewing angle or projection distorts the landmass so that Scotland is foreshortened, and the northern islands such as Orkney and Shetland are squeezed into obscurity (if they are not cut off entirely). In summary, centres and edges are made and reproduced through maps – choose the centre of your map differently, and a different world with a different politics is made. It therefore makes an enormous difference where marine energy environmental resources are located on the map, for with that mapped location comes particular politics and power.

Our second version of the centre–periphery dichotomy revolves around the urban metropolis with rural areas reduced to peripheries that must play catchup.

This model is most notably perpetuated through Manuel Castells' account of the modern Network Society.[8] This model rests upon a technological map of how information (email, Internet, bitcoin) moves around the world through communication infrastructures. It is reproduced through endless, dramatic visualisations of the networked world, with cities as hubs reaching out to other urban centres and to those of us who live elsewhere either in the dark or simply invisible. This discourse is pervasive: rural places and people are assumed to be lacking technological progress, as though technology development is a single smooth line on which we can all be located, and urban places and people are considered ahead on that same line of technological development. Many researchers have troubled this dichotomy. For example, Anna Tsing's work on *friction* shows the generative tension between urban centres and rural sites of production necessary to create globalisation.[9] Tsing's work shows how much is happening in the places described by others as "the end of the world"; such places are where global connections are established; for example, science and technology studies scholar Nicole Starosielski has conducted a study of the undersea cable networks that carry telephone and Internet data around the world.[10] She notes how its physical geography in no way correlates to the public imaginaries of this global network. For example, large data centres are not located in cities but are more often located close to water-cooling resources and renewable energy that, as we have already discussed, are more often found in high-energy, stormy edge places. So, if information is power, such power is at the geographical edge.[11]

From this initial foray into centres and peripheries, it would seem that both EMEC in the Orkney islands and DanWEC in north-western Jutland, share a similar geography and challenge. First, DanWEC in north-western Denmark and EMEC in northern Scotland are both located at the edges of their national maps. And this geography is repeated for many marine energy test sites around the world. Marine energy would appear to be located at the periphery of the map and far from powerful centres, where parliamentary political decisions are made. Second, wave and tide energy devices are, without doubt, advanced technology. Yet they are to be found, installed within their environmental resource, in stormy coastal locations that are often distant from the calm, un-weathered urban centres. If we were to follow the rigid binary of centre versus periphery, it would appear that marine energy technology, located at the weathered edge of the land, has to work against the grain of the 'technologically backwards' rural periphery.

But seen through a more careful and critical lens, marine energy devices and their infrastructures are at odds with this standard version of the centre–periphery relation. This is where ethnographic fieldwork and attention can provide a richer and more complex empirical account.

5 From periphery to edge

We stand at what some think of as the end of the world. We are in Orkney, at the geographic periphery of Europe, an archipelago of 20 or so inhabited islands, a wild mix of farmland and Atlantic wind and waves. We stand at the periphery, yet if we drove half an hour to the other side of the main island we would be at the centre of the islands, in the main cathedral town of Kirkwall, and there is much grumbling about centralisation of the islands' administration in the town.

We stand on a beach with fossil-filled sandstone rising around us, the Sun low and cool across a winter-green sea. If we squint, we can just see four red lights blinking in and out of the waves. These are the marker buoys for the European Marine Energy Centre wave energy test site. Out there, floating on the rolling horizon, we might be lucky enough to see bright yellow generators, such as the gyroscope system that is Well Oy's Penguin. Out there, in those green waves, was where the Pelamis became the world's first grid-connected wave energy machine. Behind us, under a grass-covered mound and half buried in the cliff side are the substation and equipment housing for the European Marine Energy Centre wave energy test site.

The former first minister of Scotland called this pounding of green sea and waves, 'the Saudi Arabia of marine power'. This place, this beach, is central to stories of both political and electrical power. So we stand at both the centre and the periphery; centre and periphery are fractal. You might look at a map of the UK and see London as the innovation centre, and Orkney as the technological backwater. But zoom in to Orkney, and you see the EMEC wave energy test site as the innovation centre and the outer islands in the archipelago with their limited broadband connection as the periphery. So if it is centres and peripheries' all the way down, how to consider this dichotomy otherwise? We were often told how hard the islanders have to work to resist a very real experience of being marginalised. As one environmental consultant said: "[I]t's like hyperthermia, all the blood goes to the centre. . . [W]e have to look after ourselves". Orkney is distant from London in so many ways that matter. How to keep the empirical evidence of fractal centres and peripheries in view, all at once? Where to locate ourselves?

We are now standing on a beach outside the port of Hanstholm. From here we have a perfect view to the alien construction that makes experiments in wave energy – the massive white block and four pillars that raise the Wavestar prototype out of the sea. Here, wind and waves are strong, and technologies for marine renewables that may help ensure fossil-free futures can be tested. We may say that it is an innovation centre, but it is not a place that most people would think of as a place for future making. This is definitely not Silicon Valley. But is it the periphery?

Jens Ole Bæhrenholdt and Brynhild Granås have argued against and sought to deconstruct the notion of the periphery.[12] Undoing the centre–periphery dichotomy is important, but it is also very difficult since people living in places regarded as remote often understand themselves as precisely that – peripheral. Young people, for example, when negotiating their identities, do so through binaries like future–past and centre–periphery. Should we take such binaries seriously? Are there limits to the intellectual project of deconstructing peripherality? Moving away from the binary vocabulary is one way of seeing what else is happening here at the edge, apart from repetition of what is considered centre and what periphery.

In Hanstholm, being at the 'periphery' is certainly an issue. At the bus station, you'll encounter a painted signpost with 'World's End' written on it. A manager at the Fishermen's Association spoke to us about Hanstholm as a remote area, completely invisible to Copenhagen and with limited possibilities for development. A teacher also involved in DanWEC said, memorably, that the distance from Copenhagen to Hanstholm is twice the distance from Hanstholm to Copenhagen. In the national context, Hanstholm is regarded as part of *udkanstdanmark* ('outland Denmark'), an area of limited prospects, especially for young people, around the western and southern margins of Denmark; an area also sometimes referred to as 'the rotten banana' (*den rådne banan*) – as though this periphery were a rotting skin around a more robust and healthy country.[13]

But a flyer from the region also says, 'Welcome to World's End – *a place of new beginnings*', and a popular musician, Simon Kwamm, has turned Hanstholm's historical light house into a recording studio. So being at the edge is also mobilised as a resource: peripherality is part of the branding efforts of Hanstholm. Yet framing peripheral Hanstholm as a technological cutting-edge akin to Silicon Valley, stating that this peripheral place really is a global innovation centre, does not empirically ring true. It leaves the people living in peripheral places, as well as the scholars trying to take seriously their lives and practices, at an irresolvable dead end. We must hold together both the empirical evidence of the periphery as both problem and resource and our need to move beyond the centre–periphery conceptualisation, which is so prevalent.

One way of doing this is by opening up our understanding of places at the edge. The edge marks a place of transformation, in our case where sea power is transformed into power on land. It is a boundary place where you encounter difference and mixture. But the edge is not a line on a map. It is an indication of places that share a particular experience – there is something shared between Hanstholm and Orkney, for example. Being located at the edge, in this sense of an edge experience, highlights how innovation in marine energy differs from innovation happening in other places that are more usually associated with tech industries (such as Silicon Valley).

6 Connecting the edges

There is much we can say about innovation 'at the edge' in marine energy. To do so, we must first constitute the edge analytically, drawing on our empirical evidence. We must develop the concept of an edge that is not simply one side of a centre–periphery.

Anthropologist Anna Tsing argues that edges are often zones of unpredictability on the margins of stability,[14] whether that be unstable coastline or unstable as a concept. She also argues that 'edges' are good to think with: they generate questions that counter the assumption that the margin and the marginal are in opposition to progress and development.[15] We follow Tsing in thinking through edge as a way to counter a simplistic understanding of marine energy as marginal.

What follows are three different analytic versions of the edge that can reconceptualise marine energy innovation.

6.1 Resilient edge

Peripheries are often considered as places that are vulnerable and at risk. Seen through the concept of edge, as we will show, the risky places of marine energy emerge as advantageous.

Thy, the municipality in which Hanstholm is located, has more than 200 wind turbines, Denmark's first power plant based on geothermal energy, a national test centre for large-scale wind turbines, a centre for research and development of energy technologies, and a centre for wave energy technology. Also, it is the home of (a part of) Denmark's first national park reaching across the dunes and the coastal landscape stretching inland towards the fjords. Thy brands itself as a 'climate municipality', highlighting in its branding efforts that 100% of the electricity and 80% of the heating there come from renewable sources. By this framing, it seeks to position itself as an example for national attempts at moving from a society based on fossil fuels to one independent of oil and gas.[16]

Thy municipality works to tell a counter-story to its marginalisation. Its story is one of *new beginnings*, that the municipality has a long history of being on the renewable energy front line, of being a landscape for research and development – a place where the future arrives first and is tested. The testing of wave energy is positioned as a continuation of this ongoing story. Wave energy is tamed, made familiar and enrolled in the regional story of being a 'climate municipality'.

Over the North Sea, in Orkney, this move to retell the edge as a place of new beginnings is also ongoing. Orkney calls itself a Living Laboratory, with a long history of testing energy futures. This history begins in the 1950s with the testing of an early wind turbine, through to the UK's large-scale wind energy test site in the 1980s, and on to the country's first smart grid, 700 or so micro wind turbines, and a test site for electric cars. The European Marine Energy Centre is just one more test site, the continuation of a well established trend.

In both places, marine energy is entangled in histories of an edge that is a site of technical experimentation and ongoing energy future making. The marine energy test sites are not located just due to the local environmental resource, the energetic sea. They are also entangled in wider relations: regional commitments to renewable energy development and a cultural history that emphasises the locale as a 'natural' site for technical experimentation, very different from stories about the edge told elsewhere.

We do not hear this as the edge claiming centrality. As we argued previously, marginalisation (marked by young people leaving the area due to limited job prospects) remains a serious problem. We hear this as a clear articulation that being at the edge is not *ipso facto* a negative; rather, there are well established advantages to being at the edge for scientific and technical development, and this has long been the case.

This should not be surprising. For those who live and work at similar edges, it will resonate. Island studies, along with anthropology and geography, have long argued against remoteness as a universal negative, that creativity and improvisation thrive at the edge.[17] Empirical evidence continues to show how islands, for example, are the harbingers of the future: their fine-tuned ecosystems and fragile socio-economics make them global barometers for planetary change.[18] It was the Alliance of Small Island States who began the climate change mitigation campaign, 1.5 to Stay Alive. Many of these low-lying island nations are on the front line of human-induced climate change. These edges are prophetic for the rest of the world. Rather than being behind, they are ahead – already living in the future. And it is important to remember that geographic understandings of islands are not limited to isolation by the sea but include wider forms of isolation; edges are not always wholly maritime.

In a reflection on a climate-changed future, the UK Government wrote, "[A] high-carbon world is one with more extreme weather, where we and our children are faced with the costs of adapting the way we live and the infrastructure and systems that support us".[19] Extreme weather is a given in many sites for OE. The high-energy sea, which makes these sites suitable for high-energy wave and tide generation, means serious weather and storms. Ocean energy edges are by definition often located in places where the weather can be ferocious: in Orkney, waves can be coming over cliffs in great booming explosions of salt and sea spray. Orkney and Hanstholm both face the Atlantic and its wrath.

Aside from being a sensitive barometer, the high-energy seas and weather can also lead to another edge effect. This relates to the UK Government's correlation between weather and infrastructure. Whether it is transport, telecommunications or electricity networks, infrastructure is often vulnerable in high-energy environments. In Orkney, the so-called lifeline ferry can stop for several days during winter storms, meaning that fresh food runs out at the supermarket. The electricity cables strung up on poles over the islands can be downed due to fierce winds, and the lights and phones can go out.

This has two implications. First, infrastructure has been regarded as an indicator of modernity, and its loss synonymous with a sudden and threatening loss of modern civilisation.[20] However, in high-energy places where infrastructure is unreliable, their breakdown does not lead to panic, social media meltdown or any perceived loss of civilisation. Rather, the breakdown of infrastructure networks such as food transportation or electricity is taken as business-as-usual in a modern world. In Orkney, locals reach into the chest freezer or into the bag of potatoes, and oil stoves and petrol generators are common. Such edge places are resilient to modern infrastructure breakdowns.

Second, as has been well discussed in infrastructure studies, when infrastructure breaks down, it becomes visible.[21] Thus, in high-energy edge locations, modern infrastructures, such as electricity, are visible networks and well known to those who live with and without them. Electricity does not come out of the electricity socket in the wall in Orkney or Hanstholm. It comes from the hundreds of local wind turbines or from the nearby geothermal plant (the case in Hanstholm).

Marine energy is thus being tested in places where it is visible as an extended part of the electricity infrastructure. It should be no surprise that regular tours around Wavestar were offered to the public (initiated as part of COP15 events) – and it became a known tourist attraction. In Orkney, there is a well established marine energy public: islanders have an informed opinion about wave and tide energy, made through extended reporting in the local newspaper and radio, ongoing since EMEC began ten years ago.[22]

Overall, marine energy is entangled in edge places that are resilient. We mean that in two ways. Following our argument, we mean that high-energy environments are resilient to infrastructure breakdown, leading to both mitigation experience and visible public participation in infrastructure – and thus public awareness and participation in marine energy. But we also want to draw on our prior comments about the potential for technical experimentation, which is another long held quality of the edge – our sites in Denmark and Scotland have long histories of technical experimentation. This enduring experimentation is also a form of resilience.

In Hanstholm, the teacher involved in DanWEC quoted previously told us about a local saying: "[I]s it possible? No? So we do it anyway!" We heard the history of this harbour town told as one of self-determination. The harbour walls, a huge concrete installation, were regarded as an engineering impossibility during previous centuries. Yet the people persevered and made the technically impossible happen. The Nazi occupation forced locals to leave; those who returned to these windblown sand dunes were "Klondike types . . . who took chances".

It is a similar story in Orkney. The islands' smart grid is under massive strain due to increasing production of local renewable energy. The local energy operator has issued a moratorium on further renewable energy generation, which would seem to make more wind or wave energy generation impossible. In response, the islanders have gone ahead and bought electric cars to increase capacity and are developing their own hydrogen fuel network to transform the renewable energy into another form. They persevere and make the technically impossible happen.

Resilience is an empirical experience for us as ethnographers. It is less a measure or prediction and more an ongoing, observable and mundane practice associated with living at the edge. Here, at the oceanic edge, the energy future is alive and kicking back.

6.2 Innovation edge

There are many different models for innovation at the edge. Indeed, innovation metaphors often include the edge: cutting edge, bleeding edge. However, rather than review the literature on innovation, we are interested in how innovation is done and discussed in practice in our OE generation sites and in what conclusions we might draw.

To learn about different attempts to move Danish wave energy forward and stabilize it as an industry, Winthereik participated in work to develop an interest organization for wave energy. The group refers to itself as the partnership, and under this partnership framework, wave energy device developers, municipality politicians, energy company representatives, suppliers of technology and materials, university-employed physicists and energy consultants meet to create a voice for Danish wave energy. This partnership is therefore one innovation practice for Danish wave energy.

The partners assemble around a five-step innovation model. The model is a classic phase model describing technological innovation as linear. It presents the development of wave energy technology as an incremental process where the technology is first small-scale and undergoing testing in a test basin; then in the second phase, it is tested in a slightly bigger scale, still in a lab. In the third phase, small prototypes are tested in inner Danish waters, for example in fjords. In the fourth phase, a device is tested in the sea at Hanstholm, and, finally, in the fifth phase, the wave energy goes commercial, and devices operate in offshore wave energy parks consisting of several units. During meetings, the model featured as a road map towards a future, where a small and vaguely defined community could go commercial and acquire the characteristics of an industry.

The model is both universalizing and very specific with regard to the places of wave energy innovation. The model both embeds geographical edge (Hanstholm) and the cutting edge (future). In the model, there is convergence between the two. The seascape at the edge is enrolled into the innovation model. In essence, innovation and the energy future are a matter of the wave energy generator physically moving to the edge. There is a doubling of the edge as both future and geography in the innovation model.

There is another important aspect to this partnership model: the underlying histories of renewable energy innovation and its edges in Denmark.

At one meeting, they invited one of wind power's 'grand old men' to speak and inspire them. In a spotless white shirt, he was living proof of wind's commercial success. He told how the capacity of windmills went from 30 kW to 6 MW in only 30 years. He narrated this rapid sectorial development as a combination of "small steps" and a willingness among a small group of pioneers to experiment and team up. The proposition is that the wave energy inventors should experiment together and become "Wave", which would allow them to replicate the success of "Wind".

The origins of wind energy (at the centre of Danish energy policy) are entangled in the same north-western region as the Danish Wave Energy Center; down the coast is where Vestas began, started by a local blacksmith. The origin myth and success of Wind, a story located here at the edge of Denmark, provide the model for innovation in Danish wave energy and underpins their five-step strategy. The steps assume a singular technological development path and that developers should cooperate to constitute a singular organization, akin to the politically powerful Danish Wind Energy Association. It could be said that Wind's approach is to centralise wave energy – to make it look (at least, politically) like them, as a central national energy industry. The fact that wind managed to make the move from the peripheral Atlantic west coast to central Copenhagen is part of the lure.

This is, indeed, the approach that the Danish Government has taken: with support for the creation of the partnership as the interest organization for the wave energy sector. So, in Denmark, we have an approach that assumes relocating from edge to centre as an innovation model.

In Orkney, understanding OE innovation by comparison with other tech industries is also a tactic. Neil Kermode, director of the European Marine Energy Centre, has often likened the development of marine energy to the beginnings of flight. In 2008 he said, "We are where the Wright brothers were after their first flight. We have proven we can do this, but we have not yet mastered our art. . . [It's like the first planes] are for sale but it's a long way from a commercial airline industry". This analogy reminds us that the development of OE is not just a matter of scaling up a single generator. You need more than just an iconic flying machine to go from the Wright brother's first flight to a Jumbo Jet on the runway at an airport. You also need the OE equivalent of passenger terminals, air traffic control, airline fuel and much else. In short, you also need infrastructure. This is not unsurprising given that EMEC is a test site that specialises in developing this infrastructure, somewhat differently from the inventors and engineers of wave energy generators involved in the Danish partnership. But what Neil Kermode at EMEC emphasises is an innovation model that is more than machine orientated, scaling up from basin to open sea, but that includes the end-to-end industry value chain. It is a reminder that OE innovation, to become the equivalent of a commercial airline industry, must include the whole environment and associated infrastructural support *at sea*. As with the partnership model, this innovation ends with salty, sea-cracked commercialisation. As Neil Kermode commented at another time, what they do is practical and inseparable from the sheer physicality of the sea: "You want to know how to anchor your device in eight knots of tide? It's practical. You try this, try that . . ."

The argument we are making is that innovation in marine energy is at the edge, not through some cutting-edge hubris but because OE at the geographic edge, the sea, must be inside innovation for it to be possible. In the case of the Danish partnership, the sea flows in one stage at a time through the innovation model. The surrounding water for the wave energy device gets ever more salty as they get closer to commercialisation, from a basin in the lab, to brackish water in a fjord, to the open Atlantic Ocean. At EMEC, they are committed to open sea testing for the whole end-to-end infrastructure of the prototype, from anchoring to cabling to support vessels.

This may seem something of an obvious point: commercial marine energy must include the sea. The sea edge must be inside OE for it to reach commercialisation. We make this point clear for two reasons. First, because we believe that the 'innovation edge' in OE should not be regarded as a flippant repetition of 'cutting edge' and 'bleeding edge' often used in technology development. The salty sea is a powerful force that alters what innovation is possible (and we will explore that in more detail in our final section). Second, this troubles the assumption that OE needs to become more central as a sector. As this book has emphasised, the resource for wave and tide energy is highly located and not general over the ocean. We have already discussed how high-energy seas are often a prerequisite for OE testing and deployment – which is not every coastline around the world. Indeed, there is much debate about where good environmental resources are to be found. To be blunt, the possible sites for profitable OE are much more limited compared to wind energy. The appropriate coastal edge is a limited resource compared to the breezy land it contains. The strength of OE is that it lies at the edge and will always do so. No market remodelling or tariff shift can alter the movement of the sea or the orbit of the Moon – wave and tide energy are immovable. Moreover, as we were often reminded, "[Y]ou can't just send a person out with a pot of paint and a spanner" to maintain an OE device in the sea, as you would a wind turbine. Instead, it involves the overhead costs of a boat, crew, divers and undersea equipment. Thus, we wonder if an innovation model that attempts to replicate wind energy or other mature land-based renewable energy industry is appropriate. We remain unconvinced that centralisation is an approach that will be effective. OE will always be innovation at the edge.

6.3 Edge of the wave

Our final version of the edge as analytic tool for marine energy returns to our description of Winthereik standing on the beach at Hanstholm looking towards Wavestar and of Watts standing on the beach in Orkney looking out over the EMEC wave energy test site. As we stand at our respective edges and watch the sea, we both reflect on so-called wave capture in marine energy – the very material process by which wave energy is transduced into electrical energy.

Winthereik says:

> I once tried to capture a wave. As part of my participation in a workshop, I had been assigned the task of bringing an object from my fieldwork site to the workshop. I decided to capture a wave to mimic the practices of 'wave capture'. With studied naivety and a jam jar I waded into the ocean and waited for a wave to let itself be captured. Once a wave had rolled into the glass, I screwed the lit back on, waded back in and put the wave in my backpack. Compared to the stories of 'wave capture' I had heard, this was too easy. Before walking back to the summerhouse I tried to prolong the event and sat for a while watching the waves . . . When I arrived at the workshop I realized that indeed I had not been able to capture a wave, far less wave energy. I had stopped the movement of water and hence its energies.

Looking out towards the same Atlantic but from the Orkney coast, Watts is beach-combing, looking for pieces of flotsam and jetsam that the waves have caught and thrown to shore. She finds a large pale, spherical float – perhaps it once hung over the side of a boat. Its pale glow in the Sun is marred by grey patches and a pitted surface, making it look like a full Moon rising over the beach. She places it on a stone on the beach and takes a photograph to capture the momentary installation. She names it: Found Full Moon. Today the waves lap at her feet. Tomorrow the waves might be 10 m high, making the beach impassable but leaving other pieces of floating detritus to be found another day.

The sea that OE operates in is hard to characterise and 'capture'. Indeed, hydrodynamic modelling of the sea and its waves and tides is a calculated craft. Our two moments may seem outliers in the material experience of OE, but we include them to emphasise that our two accounts are also, quite literally, accounts of OE – the power in the sea. We tell them to make present the sea as material agency that cannot be wholly modelled or reproduced by any empirical account or calculation. The sea cannot be tamed. It can only be approximated, and only in some small part. As those working in the industry have repeatedly told us, freak waves, unexpected and unpredictable waves, are potentially lethal to OE generators, whose design parameters must always be bounded and limited to operation in only certain kinds of seas.

Neil Kermode at EMEC once said that "the sea is a biologically active electrolyte filled with grit". Despite this attention to the grit and critters that are the sea, OE is awash in hydrodynamic models of the sea as water. These may give initial purchase on this slippery, wet environment, but we want to argue for the importance of retaining all the grit and biological critters that can never be contained in such models. We want to keep in view the materiality of the edge as essential to the power of OE. We tell our two unusual accounts of the edge to emphasise the flotsam and stones that will be hammering down upon a generator, just as they hammer down and land on a beach; to emphasise that, to completely capture a wave and cut it out of the sea, as in a jar, is to remove all its energy and power. To lose the materiality of OE, to lose a gritty understanding of the sea is to write out all the practical work needed to make any device operate and thus to write out its future and existence.

We want to show how this overflowing, turbulent and gritty sea is not a problem but an edge that can be ridden forward into the future. To do that work, we want to call attention to what happens in OE here at the edge; we want to call attention to that energy transformative process, transduction.

Transduction is when energy changes form. For example, when the kinetic energy in the movement of the waves or tides is transduced through a mechanism (such as underwater turbine blades) into electrical energy. Transduction makes an edge; it makes a boundary between two different types of energy. But transduction requires a transducer, some kind of agency and mechanism, and that means transduction is inherently lossy. You can never perfectly transduce all the energy in a wave or in the tide. OE is lossy, partial – it can take only part of the energy from the sea.[23] Transduction reminds us that edges are not where things end. Rather, edges are where things change form: land to sea, Moon to moving water, OE to water in a jar. Transduction as not just an engineering phenomena but can also be an analytic approach that gives us a way to float in the seas of OE, without trying to capture them whole.

Debates in OE have sometimes focused on which device should 'win out' over the others. We have seen elegant spreadsheets that compare the cost per kilowatt-hour of different generators. This is a universalising question, since it attempts to take one device modelled for one sea, as though the ocean were the same the planet over. The spreadsheet is akin to capturing a wave in a jar. Instead, the seas in which OE operates are edges all the way up and down, and so the practical experience of the actual sea involved, which cannot be known in advance, remains essential to the OE as a sector. Rather than trying to write this out as a problem to be solved, we are interested in celebrating and riding along on the overflowing, unpredictable and rock-strewn sea. The sea is an edge, where things change form, rise and fall, in ways that will always overflow our capacity to know in advance. OE may never be wholly predictable, but it will always be experiential – a gritty edge the industry can ride and transduce into electricity.

7 Conclusion

We began this account by pointing to how our industry colleagues in marine energy were already bringing together various geographical edges during our initial workshop. We interpreted their efforts as a call for remaking the old dichotomy of centre and periphery. In this chapter, we have emphasised how the usual approach to centres and peripheries as a dichotomy is problematic. Centres and peripheries are both highly contingent (they are not always the same for everyone), and, despite being mobilised in the industry, they do not serve marine energy innovation well: our two test sites and national industries are both technologically advanced cutting-edge practices and located on geographical edges.

To reflect on the serious issues of marginalisation in OE but to not be limited by that understanding, we have instead focused on the concept and experience of edge. Through our ethnographic fieldwork around the Danish wave energy industry and the European Marine Energy Centre in Orkney, we developed three different approaches to the edge specific to OE.

First, we named one edge 'the resilient edge' and described how OE seems to be located in places with long histories of technical experimentation and with highly visible electricity infrastructure – either through infrastructure breakdown or ongoing commitments to local renewable energy generation. The resilient edge appears to be living in the future for low-carbon infrastructure, which could be a tactical advantage for OE.

Second, we explored the marine energy innovation models discussed at our field sites. We noted how both marine and wind energy is entangled in these models of how OE could be harnessed in the future. We discussed how the sea, an edge essential to OE, cannot be written out of the innovation process, thus making notions of centralising OE, as land-based wind has been centralised, deeply problematic.

Finally, we explored the irrepressible materiality of the sea as an edge. We discussed how struggles to model or to capture in a computer the sea with all the flotsam and jetsam it contains does not need to be only a problem for OE. Instead, we suggest that the overflowing, physical quality of the sea is both the environmental resource that OE depends on and the physical differentiator that makes testing in local sea conditions intrinsic to the sector: marine energy innovation will always be located at the edge, and will always overflow predictions. OE is, in essence, uncontainable and demands constant adaptation to new seas where it is deployed. In essence, we see OE on an ever moving wave of innovation.

Our chapter has attempted to consider marine energy 'edges' laterally, slantwise and against the grain. We have considered the edge as both a located experience and as a concept. We do not regard these as tangential. Rather, this chapter is about how the edge is always both material and semiotic, both raging sea and raw model.

In the marine energy sector, we understand this multifaceted edge as a way to open up how we understand the industry. We find that our three versions of the edge provide a more accurate account and grasp on marine energy and its processes and sites of innovation. Rather than the edge as a lack and limitation to be overcome, we regard the edge as an inseparable part of what makes wave and tide energy devices work, now and into the future – a defining characteristic that is built-in. If we understand wave and tide energy generation as more than a technical endeavour and instead as the development of a social, cultural, environmental, political and technical infrastructure (as we have argued), then the edge becomes an important part of OE infrastructure.

To conclude, the edge is not the end of the world but is where things (whether waves or innovative ideas) transform – a zone of unpredictability that always overflows. As they glint and undulate in the always shifting waves, OE devices, along with all the people, places and work needed to make and maintain them, are not located at an edge that is falling off the map but on an edge that is always connecting, transducing, innovating, never still or powerless.

Notes

1 'Reconfigure' is more than reconceptualise. The latter assumes that a concept is semiotic and 'in the mind' (so to speak), whereas the former, reconfigure, includes the way a concept is made through practice and heterogeneous relations, and so it is both semiotic and material, not just 'in the mind'. For a discussion, see Haraway, D. (1994) A Game of Cat's Cradle: Science Studies, Feminist Theory, Cultural Studies. *Configurations*, 2(1), pp. 59–71.

2 For a discussion of ethnography as method, see Forsythe, D. (1999) 'It's Just a Matter of Common Sense': Ethnography as Invisible Work. *Computer Supported Cooperative Work (CSCW)*, 8(1–2), pp. 127–145; Marcus, G. (1995) Ethnography in/of the World System: The Emergence of Multi-Sited Ethnography. *Annual Review of Anthropology*, 24(1), pp. 95–117.

3 Star, S. L. (1999) The Ethnography of Infrastructures. *American Behavioral Scientists*, 43(3), pp. 377–391; Edwards, P. (2003) Infrastructure and Modernity: Force, Time, and Social Organization in the History of Sociotechnical Systems. In: Misa, T. J., Brey, P., and Feenberg, A. (eds.) *Modernity and Technology*. Cambridge: MIT Press,

pp. 185–225; Edwards, P. (2013) *A Vast Machine: Computer Models, Climate Data, and the Politics of Global Warming.* Cambridge: MIT Press; Harvey, P., Jensen, C. B., and Morita, A. (2016) *Infrastructures and Social Complexity: A Companion.* Abingdon: Routledge.

4 National Infrastructure Commission (2014) *Smart Power.* Available at: www.gov. uk/government/uploads/system/uploads/attachment_data/file/505218/IC_Energy_ Report_web.pdf

5 Crenshaw, K. (1991) Mapping the Margins: Intersectionality, Identity Politics, and Violence Against Women of Color. *Stanford Law Review*, 43(6), pp. 1241–1299; Hooks, B. (1984) *Feminist Theory: From Margin to Center.* New York: Routledge; Kearney, M. (1995) The Local and the Global: The Anthropology of Globalization and Transnationalism. *Annual Review of Anthropology*, 24(1), pp. 547–565; Massey, D. (1994) *Space, Place and Gender.* Cambridge: Polity Press.

6 Castells, M. (1996) *The Rise of the Network Society, Volume 1: The Information Age: Economy, Society and Culture.* Oxford: Blackwell.

7 Cosgrove, D. E. (1999) *Mappings.* London: Reaktion Books; Turnbull, D. (1994) *Maps Are Territories: Science Is an Atlas.* Chicago: University of Chicago Press.

8 Castells, M. (1996) n. 6.

9 Tsing, A. L. (2004) *Friction: An Ethnography of Global Connection.* Princeton, NJ: Princeton University Press.

10 Starosielski, N. (2015) *The Undersea Network.* Durham, NC, and London: Duke University Press.

11 Ibid.

12 Bærenholdt, J. O., and Granås, B. (2016) *Mobility and Place: Enacting Northern European Peripheries.* London: Routledge.

13 Winther, M. B., et al. (2012) 'The Rotten Banana' Fires Back: The Story of a Danish Discourse of Inclusive Rurality in the Making. *Journal of Rural Studies*, 28(4), pp. 466–477.

14 Tsing, A. L. (1994) From the Margins. *Cultural Anthropology*, 9(3), pp. 279–297.

15 Tsing, A. L. (2015) *The Mushroom at the End of the World: On the Possibility of Life in Capitalist Ruins.* Princeton, NJ: Princeton University Press.

16 Winthereik, B. R., Maguire, J. and Watts, L. (in press). The Energy Walk: Infrastructuring the Imagination. In: Ribes, D., and Vertasi, J. (eds.) *Handbook of Digital STS.* Princeton, NJ: Princeton University Press.

17 Ingold, T., and Hallam, E. (eds.) (2007) *Creativity and Cultural Improvisation.* ASA Monograph. Oxford: Berg; Leach, J., and Wilson, L. (2014) *Subversion, Conversion, Development: Cross-Cultural Knowledge Exchange and the Politics of Design.* Cambridge: MIT Press.

18 Baldacchino, G. (2007) Introducing a World of Islands. In: Baldacchino, G. (ed.) *A World of Islands: An Island Studies Reader.* Charlottetown, Canada: Institute of Island Studies, University of Prince Edward Island, pp. 1–29.

19 Department for Energy and Climate Change (2009) *The UK Low Carbon Transition Plan Executive Summary.* London: DECC Publications, p. 18.

20 Edwards, P. (2003) n. 3.

21 Star, S. L. (1999) n. 3.

22 For a discussion of how energy publics are formed, see Marres, N. (2008) The Making of Climate Publics: Eco-homes as Material Devices of Publicity. *Distinktion*, 9(1), pp. 27–45; Marres, N. (2011) The Costs of Public Involvement: Everyday Devices of Carbon Accounting and the Materialization of Participation. *Economy and Society*, 40(4), pp. 510–533.

23 Helmreich, S. (2007) An Anthropologist Underwater: Immersive Soundscapes, Submarine Cyborgs, and Transductive Ethnography. *American Ethnologist*, 34(4), pp. 621–641.

Index